Palgrave Studies in Green Criminology

Criminologists have increasingly become involved and interested in environmental issues to the extent that the term Green Criminology is now recognised as a distinct subgenre of criminology. Within this unique area of scholarly activity, researchers consider not just harms to the environment, but also the links between green crimes and other forms of crime, including organised crime's movement into the illegal trade in wildlife or the links between domestic animal abuse and spousal abuse and more serious forms of offending such as serial killing. This series will provide a forum for new works and new ideas in green criminology for both academics and practitioners working in the field, with two primary aims: to provide contemporary theoretical and practice-based analysis of green criminology and environmental issues relating to the development of and enforcement of environmental laws, environmental criminality, policy relating to environmental harms and harms committed against non-human animals and situating environmental harms within the context of wider social harms; and to explore and debate new contemporary issues in green criminology including ecological, environmental and species justice concerns and the better integration of a green criminological approach within mainstream criminal justice. The series will reflect the range and depth of high-quality research and scholarship in this burgeoning area, combining contributions from established scholars wishing to explore new topics and recent entrants who are breaking new ground.

David R. Goyes
Editor

Green Crime in the Global South

Essays on Southern Green Criminology

palgrave
macmillan

Editor
David R. Goyes
University of Oslo
Oslo, Norway

Palgrave Studies in Green Criminology
ISBN 978-3-031-27756-6 ISBN 978-3-031-27754-2 (eBook)
https://doi.org/10.1007/978-3-031-27754-2

This Palgrave Macmillan imprint is published by the registered company Springer Nature Switzerland AG
The registered company address is: Gewerbestrasse 11, 6330 Cham, Switzerland

Preface

This book presents a socio-criminological study of environmental crime in the global South. It gathers contributors from all the regions of the geographical global South (Africa, Asia, Oceania, and Latin America) and various representatives of the metaphorical South (i.e., historically impoverished and oppressed populations) to discuss instances of environmental crime and conflict. Overall, the aim is to further decolonize the knowledge production of green criminology. What is considered is the legacy of colonisation and the North–South and core–periphery divide in the production of environmental crime, the epistemological contributions of the marginalised, impoverished, and oppressed, and the unique contexts of the global South. This book has three parts: the dynamics of environmental crime in the global South; the responses to environmental crime in the global South; and epistemological bridges between the South and the West. The first two parts represent the breadth of the topics that green criminologists have historically studied but from unique perspectives. The third part explores ethical and decolonial ways for

Southern green criminology to collaborate with Western academia. What is sought in this book is the advancement of a transnational criminology inclusive of the knowledge of the global South.

Oslo, Norway David R. Goyes

Acknowledgements

The contributors to this book acknowledge that we carry out our work on Indigenous territory. We also acknowledge that Indigenous lands sustain us culturally, economically, and socially; and that the original inhabitants and protectors of these lands have suffered from forced displacement and genocide. We endeavour to redress, to the best of our capacities, the cultural and ecological debt of humanity.

Praise for *Green Crime in the Global South*

"Colonialism, since its beginning, has committed horrible genocides. The international decolonial movement, much to the discomfort of colonial powers, insists on obtaining redress. And Southern green criminology shines a new light on crimes that were only tangentially considered before: green colonial crimes. This book offers a global overview of this type of criminality, which is no less startling than the genocides. Genocide and ecocide are not independent either—they usually interlock. The paradox, however, is that punitive power is functional in the commission of both. In this book, the reader will understand the origins of this paradox by looking at the roots and trajectories of colonial ecocide around the world. Readers will likely get anxious reading this book—but the responsibility does not lie in the book but in the facts."

—Eugenio Raúl Zaffaroni, *Former judge of the Interamerican Court of Human Rights*

"Many of us who write about corporate and state crime are neglectful of the legacies of colonialism. *Green Crime in the Global South* is a fine collection, edited by Colombian scholar David R. Goyes, that corrects this neglect. Overwhelmingly written by scholars from the global South, this beautifully integrated book takes green criminology to a new place with a searing emphasis on state corporate crime. It weaves together a thoughtful plurality of perspectives from all corners of the planet. Goyes has curated a splendid conversation for the planet that positions Southern green criminology as transcending colonial legacies."

—John Braithwaite, *Australian National University and University of Maryland, USA*

"Those who seek justice, equality, and freedom are not voiceless. They are not powerless. Their agency, power, and voice have been selectively and skilfully tuned out by the systems that thrive on their oppression. Colonialism and extractivism have continued to inform the experiences of people and the impacts on the environment of countries in the global South. The skewed power relations between the global North and the global South have contributed to the ongoing devastation of the environment without accountability, recourse, and environmental reparations for global South nations. The essays in this book are therefore a timely contribution to these and related issues. They offer a critical reflection that not only enriches the discourse but is an amplification of our voices on the demands for change and justice."

—Busisiwe Kamolane-Kgadima, *Attorney and Activist, Johannesburg, South Africa*

Contents

Notes on Contributors

Karine Ágatha França is a Ph.D. candidate at the Federal University of Santa Catarina, in Brazil. In her masters' thesis, she studied perceptions about environmental harm in the case of the dam collapse in Brumadinho, MG, Brazil, addressing reflections on transitional justice in territories highly contaminated by the mining industry.

Avijit Chakrabarty Ayon is a lecturer in Sociology at Shahjalal University of Science and Technology, Bangladesh. His research interests focus on the environment, technology, health, and gender dynamics in Bangladesh.

Halil Ibrahim Bahar graduated from the Police Academy in 1988 in Ankara, Turkey. He received his M.A. and Ph.D. from the Department of Criminology, Leicester University, UK. From 1995 to 2015, Bahar taught sociology, criminology, and victimology at the Turkish Police Academy. From 2015 to 2016 he taught introduction to sociology, political sociology, Turkish modernization, political sociology, and introduction to political sciences at Aksaray University, Turkey. He is now a freelance researcher and analyst in Ankara. His academic interests include social control, terrorism, domestic violence, juvenile delinquency, policing policies, migration, urban security, and green crime.

María Laura Böhm is an Argentine-German lawyer UBA (Universidad de Buenos Aires) and a criminologist at both the National University of Lomas de Zamora (Argentina) and the University of Hamburg (Germany). Also in Hamburg she obtained her Ph.D. in Social Sciences and was later a postdoctoral researcher on a grant from the AvH Foundation at Georg-August-Universität Göttingen. Between 2015 and 2019 she was a fulltime DAAD (German Academic Exchange Service) Long-Term Guest Law Professor UBA (Universidad de Buenos Aires). Since August 2019 she has been researching and teaching at the Ludwig-Maximilians-Universität Munich, where she obtained her Ph.D. in Law in May 2022.

Amy Couper is a former associate lecturer and Ph.D. graduate in the School of Justice at QUT, Australia; she is currently a serving officer with the Queensland Police Service. Her research advocates for a non-speciesist criminology and focusses on the media constructions of sharks and the ways in which governments legitimate unfounded culling practices for political and profitable gain.

Bruna dos Santos L. da Silva was awarded an Emerging Leaders in the Americas Scholarship. She is a teacher assistant at Carleton University, Ottawa, Canada and a master's candidate at the Latin America Integration Program, University of São Paulo.

Marília de Nardin Budó is a professor at the School of Law at the Federal University of Santa Catarina (UFSC), Brazil. Her research focuses on crimes of the powerful, green criminology, mass media and crime, and decolonial theory. She is the author of *Mídia e Controle Social: da construção social da criminalidade à reprodução da violência estrutural*, *Mídias e discursos do poder: estratégias de legitimação do encarceramento da juventude no Brasil*, and, co-edited with D. Goyes, L. Natali, R. Sollund, and A. Brisman, *Introdução à criminologia verde: Perspectivas críticas, descoloniais e do Sul* (Tirant Brasil, 2022).

Nikhil Deb is an assistant professor of Sociology at California Polytechnic State University, San Luis Obispo. His research focuses on how political and economic activities typically favour some human groups at the expense of other existing or future humans or the environment, with particular attention to agrarian communities in Bangladesh and India. His publications have appeared in several peer-reviewed journals and edited volumes, such as *Social Problems* and *Critical Sociology*. He is currently co-editing a book titled *Social Justice in the Global South* (Edward Elgar).

José Luis Carpio Domínguez is a doctor in the Social Sciences with a focus on Sustainable Development from the Universidad Autónoma de Nuevo León. He is a research professor at the Unidad Académica Multidisciplinaria Reynosa Aztlán of the Universidad Autónoma de Tamaulipas (UAMRA-UAT). He is a member of the Mexican Society of Criminology (SOMECRIM) and the Thematic Network of Forensic Sciences of CONACyT. He is conducting scientific research in green criminology with topics related to different forms of environmental crime, socio-environmental conflicts, and the interrelation between public safety and the environment.

David R. Goyes is a senior researcher in the Department of Criminology and Sociology of Law, University of Oslo, Norway. He earned his law degree from the National University of Colombia in 2009 and his Ph.D. in criminology from the University of Oslo in 2018. From 2012 to 2014, he worked as a rapporteur in the Colombian peace negotiations with the FARC guerrillas. He is a pioneer of green criminology in Latin America, first proposing it in 2012, and in 2019 he published *Southern Green Criminology*.

Orika Komatsubara is a JSPS Research Fellow (Young Researcher). She is affiliated with Kansai University and is a visiting research associate at the Leuven Institute of Criminology, KU Leuven. She holds a Ph.D. in Human Sciences from Osaka Prefecture University in Japan. Her research focuses largely on restorative justice. She is the author of 『性暴力と修復的司法』(*Sexual Violence and Restorative Justice*) (Seibundo, 2017). She received the award of the Japan Association of

Gender and Law in 2018. Her latest publication is 'Can art convey a victim's voice to future generations? A case of Minamata disease in Japan', *The International Journal of Restorative Justice*, 5(3): 423–439.

Gabriela Gallegos Martínez is a doctor in the Social Sciences with a focus on Sustainable Development from the Universidad Autónoma de Nuevo León. She is a research professor at the Universidad de Monterrey and has experience in the research, implementation, and evaluation of social interventions with an intersectional, intercultural, and participatory perspective, particularly with vulnerable groups in rural, peri-urban, urban, and Indigenous areas. Her research lines are in Indigenous youth, intercultural processes, higher education, and urban Indigenous migration.

Dave Mankhokwe Namusanya is a doctoral candidate at Abertay University, Scotland. He conducted his research in Southern Malawi on climate change and water practices. The research used ethnographic approaches to understand the ways through which communities in Malawi are adapting to climate change impacts on water.

Lorenzo Natali is an associate professor of Criminology at the School of Law, University of Milano-Bicocca, Italy. His research focuses on violent crime, symbolic and radical interactionism, green criminology, and visual and sensory methodologies. He is the author of *A Visual Approach for Green Criminology: Exploring the Social Perception of Environmental Harm* (Palgrave, 2016), *Cosmologías violentas: Itinerarios criminológicos* (with Adolfo Ceretti) (Marcial Pons, 2016), and *Green Criminology: Prospettive Emergenti sui Crimini Ambientali* (Giappichelli, 2015). He co-edited with M. de Nardin Budó, D. Goyes, R. Sollund, and A. Brisman, *Introdução à criminologia verde: Perspectivas críticas, descoloniais e do Sul* (Tirant Brasil, 2022).

Cristina Rego de Oliveira is a postdoctoral researcher at the Faculty of Law, University of São Paulo (FRDP/Brazil). She has a Ph.D. in Law, Justice, and Citizenship in the Twenty-first Century from the Faculty of Law and Centre for Social Studies, University of Coimbra/Portugal. She obtained her master's degree in Criminal Law at the University of Coimbra/Portugal.

Daniela Arantes Prata is a Ph.D. candidate in Law at the London School of Economics. She obtained her LLM (first class honours) from the University of Cambridge, King's College. Her master's in Criminal Compliance is from University Castilla-La Mancha (UCLM); her LLB is from the University of São Paulo (FDRP-USP), Brazil.

Jesús Ignacio Castro Salazar is a doctor in the Social Sciences with a focus on Sustainable Development from the Universidad Autónoma de Nuevo León. He is a research professor at the Instituto Tecnológico Superior de Abasolo, Guanajuato in the Coordination of Environmental Engineering. He is a member of the National System of Researchers of Mexico. His main lines of research are environmental public administration and policy, sociology, and environmental legislation.

Nigel South Ph.D., FacSS, is Emeritus Professor of Sociology at the University of Essex. He is European Editor of *Critical Criminology*. In 2013 he received a Lifetime Achievement Award from the American Society of Criminology, Division on Critical Criminology and Social Justice; in 2022 he received the Outstanding Achievement Award from the British Society of Criminology. Recent projects (with David Rodriguez Goyes and others) have focused on Indigenous communities in Colombia and environmental and health justice issues. His books include, (co-edited, 2020) *The Routledge Handbook of Green Criminology* (2nd edn.) and (co-authored, 2018) *Water, Crime and Security in the Twenty-First Century*.

Reece Walters is a professor of Criminology, School of Humanities and Social Sciences, Deakin University, Australia. His areas of research interest and expertise include green and Southern criminology, crimes of the powerful, and the sociology of criminological knowledge. He is particularly interested in the ways in which corporations and governments exploit and compromise the 'essentials of life', namely air, food, and water, for power and profit.

Rob White is Emeritus Distinguished Professor of Criminology at the University of Tasmania, Australia. He has published extensively in the areas of youth studies, criminology, and eco-justice. Among his recent books are *Advanced Introduction to Applied Green Criminology* (Edward

Elgar, 2023), *Theorising Green Criminology* (Routledge, 2022), and *Critical Forensic Studies* (with Roberta Julian and Loene Howes) (Routledge, 2022). He is Editor-in-Chief of *Forensic Science International: Animals and Environments*.

List of Figures

List of Tables

1

Southern Green Criminology: Fundamental Concepts

David R. Goyes

The Global Landscape of Environmental Destruction

The *Environmental Justice Atlas*, an initiative of the Universitat Autónoma de Barcelona, maps socio-environmental conflicts around the world (Temper et al., 2015). These are conflicts defined by the atlas creators as 'mobilizations by local communities against particular economic activities whereby environmental impacts are a key element of their grievances' (pp. 261–262). As of December 2022, the global South is the location of most of the social clashes driven by environmental issues (see Fig. 1.1). While the *Environmental Justice Atlas*'s depiction may not be 100% accurate because of the intrinsic difficulties associated with reporting conflict (Coleman & Moynihan, 1996), it does witness

D. R. Goyes (✉)
University of Oslo, Oslo, Norway
e-mail: d.r.goyes@jus.uio.no

D. R. Goyes (ed.), *Green Crime in the Global South*, Palgrave Studies in Green Criminology, https://doi.org/10.1007/978-3-031-27754-2_1

to the uneven distribution of environmental conflict and victimisation across the world.

The atlas also maps the actors involved in socio-environmental clashes, thus making visible the unequal nature of ecological exchanges between the global North and the global South. Corporations from the global North are over-represented as the exploiters of Southern natural resources; local populations are the primary protectors of the ecosystems they inhabit. The ecosystems of the global South are depleted, the fundamental needs of its human and non-human populations go unmet, and the lives of those who endeavour to defend nature are threatened, while the global North benefits from the environmental resources it has plundered (Hornborg, 2009; Lessenich, 2019; Martinez-Alier, 2018).

The following are vignettes from four continents of the global South about environmental degradation that exemplify the patterns of imbalance in the ecological exchanges between the global North and the global South.

Africa: The oil company Pemex, owned by the Mexican state, exports about 80% of its crude to the USA (Reuters, 2022). A by-product of Pemex's oil production is coker gasoline, which contains large amounts

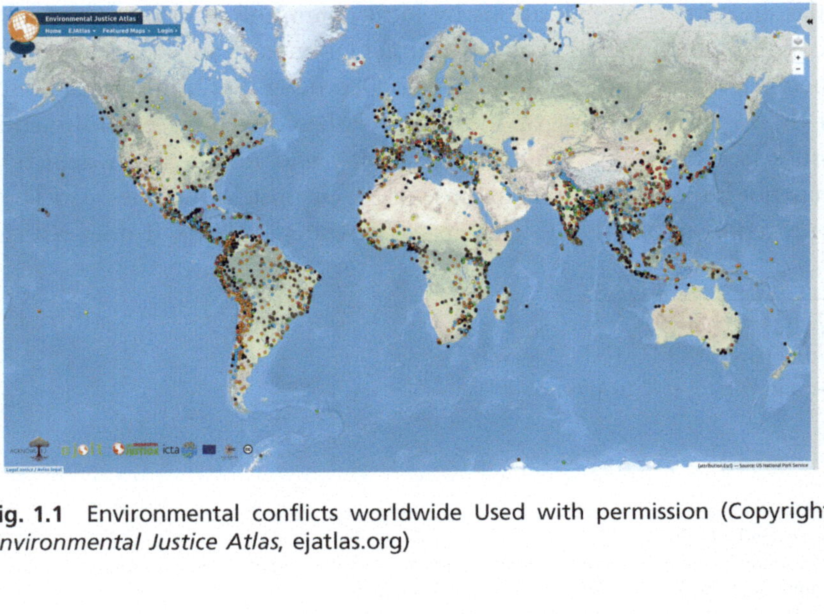

Fig. 1.1 Environmental conflicts worldwide Used with permission (Copyright: *Environmental Justice Atlas*, ejatlas.org)

of sulphur and silica—highly hazardous waste products. In 2006, Pemex sold its coker gasoline to Trafigura, a UK oil trading company, trusting that the corporation would safely discard the hazardous material. It was first transported in trucks to Texas, USA, and then loaded onto the oil tanker *Probo Koala*, flying a Panamanian flag. The ship, designed to transport oil and oil by-products, was equipped to refine the coker gasoline into naphtha, a flammable liquid hydrocarbon mixture. Mixing coker gasoline with a caustic solution, the crew successfully produced naphtha, thereby not only saving costs by avoiding the expenses of a refinery but also creating the possibility of making a profit out of the new product. (Trafigura sold the naphtha for USD19 million.) In the process, however, an even more hazardous product—banned throughout Europe—was created: sulphurous waste (Duckett, 2009). On 2 July 2006, the *Probo Koala* approached the Port of Amsterdam to discard the waste, but was unable to offload it because the port did not have the facilities to treat it (MacManus, 2016). The ship continued, and on 19 August arrived in the Port of Abidjan, the de facto capital of Ivory Coast. Here Trafigura transferred the waste to Compagnie Tommy, a local company that dumped the toxic waste in 18 locations around Abidjan (Amnesty International, 2016). At least ten Ivorians died because of the contamination and more than 30,000 sought medical attention (Franko & Goyes, 2019).

Asia: For decades, the global demand for palm oil has been increasing. In just one decade, the volume of palm oil production has increased 19% from 56 million metric tonnes in 2012 to 75.5 in 2022 (Statista, 2022). Palm oil is used in food and non-edible products (Reardon et al., 2019), but most notably the 'growing concern over diminishing crude oil reserves, fluctuating oil prices, and unease with the dependence on oil supplies … adds to the surge in interest in palm oil' as a biofuel (Mol, 2017, p. 3). Seventy-five percent of palm oil is traded on the international market, with the European Union steadily increasing its consumption (Mol, 2017). The significant demand in the global North makes 'heavy reliance on imports from the global South inevitable' (p. 4). Most of the 17 million hectares of palm oil crops in the world are located in South Asia (Mol, 2017). Indonesia, Malaysia, and Singapore now face significant socio-environmental problems caused by palm oil production. Environmental problems include 'greenhouse gas emissions from the clearance of primary tropical rainforest, habitat and biodiversity loss, water and air pollution, soil erosion, and peatland degradation' (Reardon et al., 2019, p. 118). Social problems encompass systematic breaches of workers' welfare, land grabbing, and the resulting forced displacement of local populations. As Mol (2017) stated:

> The expansion of palm oil production in the Global South, driven (amongst other factors) by the demand for renewable energy feedstock in the framework of climate change mitigation, has been critiqued as a form of 'CO2lonialism' that exploits and destroys human and nonhuman life in regions where the industry operates. (p. 4)

Latin America: Oil exploration in Ecuador began in 1879 when M. G. Mier and Company received exclusive rights to extract petroleum in the Santa Elena Peninsula. Six decades later, another North American corporation, Shell Oil Company, received the first concession in eastern Ecuador (Yeager & Smith, 2017). In 1964, Texaco Petroleum Company joined the foreign delegation in prospecting for crude upon an invitation from the Ecuadorian government that wanted to 'develop' the country (Gutierrez, 1974; Yeager & Smith, 2017). The decades-long extraction of oil gave the North American corporations and Ecuadorian elites substantial economic benefits but at the expense of the original inhabitants of oil-rich areas who were expelled from their territories and saw their ecosystems polluted. In 1990, Texaco gave up its operations in Ecuador but left behind significant environmental destruction: decimated rainforests, polluted rivers, and sick human and non-human inhabitants. The corporation's careless disposal practices left an indelible scar on nature. After prolonged litigation against Texaco, the corporation implemented a USD40 million programme to replace soil and trees and to purify the contaminated water. The Ecuadorian Ministry of Energy and Mines declared that the North American corporation had fulfilled its reparative environmental duties; however, only a fraction of the environmental damage was addressed and the human devastation was entirely ignored (Yeager & Smith, 2017).

Oceania: Approximately 10 million people inhabit the Pacific Islands. Isolated from the rest of the world by a vast ocean, they face myriad threats to their existence caused by a long history of Northern colonialism (Ramírez, 2021). Climate change, to which nations in the global North have contributed with 92% of the excess emissions (Hickel, 2020), threatens their survival as their countries are being flooded by rising ocean levels. In the Republic of the Marshall Islands, the easternmost country in Micronesia, a dome containing waste from the US nuclear programme is threatening to collapse (Rust, 2019). Human and non-human inhabitants of the Cook Islands are threatened by the pressure to mine the seabed for metallic nodules, rich in metals and minerals, needed to make batteries (Readfearn, 2022).

A list of the most salient ecological interactions between the global North and the global South, as evinced by the above examples, includes:

- The capitalist rush to increase profit at an exponential pace fuels the devastation of Southern ecosystems (Stretesky et al., 2014).
- The global thirst for resources to satisfy consumerist cultures in Northern countries takes its biggest ecological and social toll in the global South (Brisman & South, 2014; Lessenich, 2019).
- While Northern corporations are the major players in the extraction of natural resources, Southern lands are left with a landscape of environmental depletion through ecological extraction (Böhm, 2020a, 2020b; Stretesky et al., 2014).
- Northern countries do not only consume the environmental goods of the global South, but they also transfer to the South the environmental bads (i.e., waste) they generate (Ruggiero & South, 2013).
- National governments in the global South, which collude with Northern corporations, facilitate the ecological victimisation of their citizens.
- Indigenous populations have historically been the main direct victims of ecological devastation (de Carvalho et al., 2020; Goyes, South, Abaibira et al., 2021).

The North–South divide in the landscape of global environmental destruction is not random; it is the result of a historical trajectory of (neo)colonialism with identifiable actors. Intending to provide the analytical and theoretical tools to capture and dissect the North–South ecological interactions, the first explicit propositions for a *Southern green criminology* were laid out in 2016 (Goyes, 2016). This is a framework for the sociological and criminological study of environmental crimes, harms, and conflicts in the global South that indicates the legacies of colonisation, North–South and core–periphery divides, the epistemological contributions of the marginalised, impoverished, and oppressed, and the particular contexts of the global South (Goyes, 2019, 2021, 2022).

The Fundamental Concepts and Theories of Southern Green Criminology

The contemporary world is shaped by a constellation of flows of capital and surplus value, goods (including environmental resources), individuals, and information (including political influence, knowledge, and norms) that move through local, national, regional, transnational, and global contexts (Goyes, 2021). In it, myriad actors throughout the world are involved in initiating, mediating, and receiving diverse types of flows. For instance, tax havens play a significant intermediary role in environmentally degrading practices (Galaz et al., 2018). The flows navigating the globe travel through several geographical spaces—from the local to the global—generating a constellation of results, many of which are predictable and patterned, but others are unpredictable, diverging from the norm. Consequently, the study of global environmental relationships requires acknowledging the 'intensity of transnational connections, but also … paradoxes, unevenness, concrete modalities and disconnections' (Aas, 2013, p. 21). Southern green criminology makes use of diverse concepts and theories to account for the steady and homogeneous *and* for the changing and heterogeneous.

Of fundamental importance is the definition of 'Southern green criminology'. As Moosavi (2018) noted when analysing Southern criminology, it matters whether it is 'a theory, paradigm, school, project, model, approach, perspective, toolbox, method, lens or attitude' (p. 261). Dealing with the definition of green criminology, Brisman and South

(2020) declared that 'there is something of a debate about the appropriate name or label for this sub-field or perspective' (p. 4); they championed the term *perspective*, 'a loose framework or set of intellectual, empirical and political orientations toward problems (crimes, harms and offences related to the environment, different species and the planet)' and noted that 'it is both a network of interested individuals and a forum for sharing and debating ideas' (p. 40).

Southern green criminology, with its focus on North–South relations, more than a perspective is a *framework*, 'a logically developed and connected set of concepts and premises—developed from one or more theories—that a researcher creates to scaffold a study' (Varpio et al., 2020, p. 989). Southern green criminology invites a diversity of voices, mainly those traditionally excluded from academic debates, to use their critical creativity in the study of (neo)colonial harms. Yet, its focus is more precise that that of green criminology in that (1) attention is given to ecological discrimination and victimisation in the global and metaphorical South, and (2) its primary emphasis is on structures and processes that are associated with a higher intensity of environmental crimes and harms. As a framework, Southern green criminology contains a set of concepts and related theories useful in the interpretation of green crimes and harms in the South.

Concepts are 'assumptions about a phenomenon' (Johannessen et al., 2020, p. 29) or the properties that we assign to it. Among the central concepts that Southern green criminology uses are *global North* and *global South*; *centre* and *periphery*; *colonialism*; *internal colonialism*, *neo-colonialism*, and *imperialism*; and *coloniality* and *the epistemologies of the South*. *Theories* are 'assumptions about the relationships between phenomena' or the 'assumed connections and often connections of causality' (Johannessen et al., 2020, p. 30). Southern green criminology contains a series of theoretical assumptions about the interaction between North and South and its consequences in the environmental realm. The field also relies on political ecology and adjacent disciplines. Yet as Snipes et al. (2019) explain, theories are 'always a statement about probability, never an assertion of certainty' (p. 6). Consequently, while the existence of (neo)colonialist relations between the global North and the global South are likely to exist, Southern green criminology avoids a dogmatic reading of global interactions but identifies, based on empirical material, whether and when the colonial dimension exists (Goyes, 2018a).

First and foremost, Southern green criminology is dependent on the global **North–South** division as a central conceptual category (Goyes, 2019, 2021). This analytical concept is the direct result of historical global trajectories, in which the legacy of colonial times is that most colonising countries are located in the global North while most colonised countries are located in the global South (Franko & Goyes, 2019). Even though the USA was a colonised location of the 'New World', it forms part of the global North, not only geographically but also because the original inhabitants were decimated, along with their cultures, and almost fully replaced with British colonisers (Zaffaroni, 2022). This concept is invaluable in the study of 'the historical impact of colonial and imperial practices' in instances of conflict and violence (Aas, 2013; Carrington et al., 2019). This is, therefore, a 'relational categor[y]' that pays attention to 'links between sites and across time, such as historically (post)imperial trajectories when interpreting the current contours of world politics' (Haug, 2021).

Attached to the concept of the global North–South division is theorisation by Southern green criminologists about the interaction between the two entities. Based on the acknowledgement that the colonising structure of North–South remains intact, the framework indicates that colonisation left an uneven distribution of political, economic, and epistemological power in favour of the global North that allows Northern countries to:

- Employ international legal instruments that regulate human interaction with nature (in terms of both physical resources and immaterial property rights) (Goyes, 2017; Goyes & South, 2016);
- Extract and dispose of environmental products from Southern countries even at the cost of environmental destruction and conflict (Franko & Goyes, 2019);
- Impose environmental practices on the South to the detriment of local practices (Goyes, 2018b, 2019, 2020).

The daily dynamics of Northern and Southern countries are inseparably linked (Lessenich, 2019) in such a way that the implementation of global policies developed by Northern countries threatens basic collective needs in Southern countries (Böhm, 2018; Brisman & South, 2013; Goyes & South, 2016). Therefore, Southern green criminology identifies the North–South divide as a key driver of environmental conflict and crime (Escobar, 1995; Franko & Goyes, 2019; Goyes, 2019, 2021).

The world is, however, in constant flux, with emerging powers exerting increasing influence in political and economic realms (Bull, 2021). China and India have grown in recent decades to become global powers and challenge the dominance of Europe and North America, regrettably also increasing their environmental devastation. What used to be a clearer divide of global North and global South can now be better described in terms of a polycentric global order (Galaz et al., 2012). Responding to the realities of a changing world, Southern green criminology also uses the concepts of core and periphery offered by world-system analysis (Wallerstein, 2004). **Core countries** have quasi-monopolies over financial institutions, media and communication systems, technologies, and weapons of mass destruction (Amin, 1997). **Peripheral countries** generate raw materials and supply an 'unqualified' labour force that core countries exploit (Wallerstein, 2004). Due to the colonial trajectory, Northern countries often are the core countries and Southern countries the peripheral. However, changes in global geopolitics means that former colonies in the global South are emerging to challenge the core.

In the Southern green criminology theorisation of core and periphery interactions, core countries, using their monopoly of financial institutions, establish global economic rules and exchange products in a dynamic that results in a 'flow of surplus-value (meaning here a large part of the real profits from multiple local productions) to those states that have a large number of core-like processes' while impoverishing the periphery (Wallerstein, 2004, p. 18). Furthermore, peripheries in core countries—mainly Indigenous territories—are plundered by central governments (Bjørklund, 2020).

The global North and global South and core and peripheral countries have engaged in myriad interactions in the past 500 years that have given rise to another set of concepts: colonialism, neo-colonialism, imperialism, and coloniality. While they are used interchangeably in academic literature (Kumar, 2021), delineating their meaning highlights important analytical nuances for Southern green criminology. The interaction between the global North, represented by European powers, and the global South encompassing Latin America, Africa, and to an extent Asia was through colonialism (Zaffaroni, 2022). **Colonialism** is the direct physical occupation of a territory originally inhabited by other human groups—'a state's successful claim to sovereignty over a foreign land' (Mahoney, 2012, p. 2). It usually brought about the (1) decimation of Indigenous populations, (2) destruction of the natural environment, and (3) defacement of millennia-old ways of interacting with nature (Goyes, 2022). Often, colonialism creates and coexists with **internal colonialism**, 'a structure of social relations of domination and exploitation among heterogenous cultural groups within a single state' (Mondaca, 2017, p. 37). **Neo-colonialism** is the de facto occupation of a country via land grabs after the country has gained independence from its coloniser (Goyes & South, 2016) and the consequential imposition of colonial ways of interacting with nature. **Imperialism** is the indirect control of nations, not primarily as physical occupation but as economic and political manipulation (Goyes, 2022).

A significant portion of Southern green criminology focuses on epistemology (the assumed capacities of knowledge creation). In this realm, the designations *global South* and *global North* capture 'not only systemic inequalities stemming from the "colonial encounter" and the continuing reverberations of (mostly) European colonialism and imperialism but also *the potential of alternative sources of knowledge*' (Haug et al., 2021, p. 1928; italics added). While in most locations the colonial period ended in the late nineteenth century, the epistemological logics of colonialism survived in what Quijano (2000, 2007) defined as **coloniality**: the preference to consider valid only what follows the modern European way of knowledge creation (see also Dimou, 2021). Coloniality, environmentally speaking, takes the shape of what I term **ecological discrimination**: 'The systematic negative differentiation and oppression of some human groups, non-human animals and ecosystems, based on modern instrumental ideas about how to treat and relate to the natural environment' (Goyes, 2019, p. 15).

As a response to the coloniality of knowledge, Santos (2014) coined the phrase **epistemologies of the south**, or simply **the south** (in lower case), to indicate the knowledge that the discriminated against and marginalised produce and their ways of producing it. The south in this sense is metaphorical as opposed to the geographical global South and refers to the marginalised, oppressed, and impoverished; the south refers to the peripheral voices wherever they are located, even in the global North (Goyes, 2016, 2019). Santos's use of epistemology is unusual but reveals the theoretical implications of the concept. For him, epistemologies are a rhetorical tool that signal both method and validity and highlight the validity of the production of knowledge in the daily struggle to survive of the marginalised, impoverished, and oppressed (Goyes, 2016). Throughout the years, the epistemologies of the South have proven to be not only valid but also important in 'the development of a green criminology that contemplates responses to environmental degradation and destruction': They allow for a deeper and better contextualised knowledge of the issues, they cohere with criminology's original goal to produce knowledge about the events that affect the lives of citizens, and they reassert the validity of plural ways of knowing (Goyes, 2016, p. 514).

Since 2016, when the first explicit proposition for a Southern green criminology came about, scholars have applied its concepts and theories to colonial-driven green crime throughout the world (Arroyo-Quiroz & Wyatt, 2018; Böhm, 2018; Budó et al., 2022; Goyes et al., 2017, 2019, 2022; Goyes, South, Sollund et al., 2021; Melchiors, 2022). *Green crime in the global South: Essays on Southern green criminology* is another step forward in the construction of a conceptual framework that allows us to understand deeper and better green victimisation in the South. The contributors bring about a wealth of empirical material, theoretical insights, and refreshing voices. They write from and about Africa, Asia, Latin America, Oceania, and Europe.

About This Book

This book seeks to provide the reader with the advances made in the field of Southern green criminology around the world. Not only will this book inform readers about the global events that result in ecological destruction in Southern locations and update them on the cutting-edge theoretical innovations that explain the dynamics behind the phenomenon, but the collection also serves as a meta-study that gathers the work of academics and identifies the patterns of destruction, resistance, and knowledge creation about environmental conflicts in the global South. Inequalities in funding for research is a primary cause of the *assumed* epistemological imbalance between the global North and global South (Briggs, 2023), as it is for research about green crime. (The fact that scholars from the global North can afford to conduct macro-projects gives the incorrect impression that they are the ones producing most knowledge.) Additionally, there is a preference for mega-studies and big data (Dean, 2017; 'Is science only for the rich?', 2016; Smithers et al., 2016) from which most researchers in the global South are excluded because of limited funding in their countries. Southern green criminologists, however, address those limitations creatively (de Carvalho et al., 2020) and in solidarity. The collective case studies in this book serve as an expansive study of the dynamics of colonial environmental exploration of the global South and demonstrate that ecological discrimination and colonialism are not exceptional instances but the rule in global interactions.

The chapters of this book are divided into three parts. Part I, 'Drivers of Green Crime in the Global South', focuses on specific global dynamics that escalate environmental destruction and victimisation in the global South. Part II, 'Responses to Environmental Crime in the Global South', gains inspiration from the epistemologies of the south and the particularities of Southern contexts to propose better ways to redress green crime and harm than those offered by contemporary hegemonic ways (Goyes, 2020). The chapters in Part III, 'Global Dialogues About Crime and Destruction in the South', reflect on why and how scholars from the global South can productively cooperate with academics based in the global North to redress the effects of colonialism and curb global environmental harm. What follows is a summary of the chapters and a discussion of the key contribution each chapter makes to Southern green criminology.

Drivers of Green Crime in the Global South

In Chapter 2, 'The State-Corporate Crime of Extractive Industries', María Laura Böhm presents a theoretical framework—*raw-materials-related state-corporate crime*—substantiated by extensive empirical material, to understand the crimes of extractive industries in the global South and account for the widespread collusion between governments in the global South and Northern corporations. The corporations in the equation are transnational companies that exploit the relative weakness of states in the global South to profit from resource extraction. This means that political elites fail to hinder corporate harms and even encourage corporations to develop extractive activities in Southern soil. The collusion of corporations and states in pursuit of their respective interests—which often coincide—result in socially and environmentally injurious state-corporate crime. As both this introduction and Böhm's chapter reveal, environmental victimisation in Southern locations is strongly associated with the presence of transnational corporations, and although the visible executors of the harm are locals, transnational actors are the most important (although invisible) forces behind environmental degradation. These corporations exploit the narrative that Southern countries can 'develop' by offering their environmental resources to Northern economies. Böhm calls this phenomenon *economic opportunism*.

Böhm's framework reveals the social structures sustaining systematic environmental harm, providing readers with tools to conduct analyses of state-corporate crime involving raw materials. She assigns six traits to raw-materials, state-corporate crime: *actor scattering*, the diffusion of actors around the world in shifting positions; *temporal and spatial complexity*, the many time and space factors that facilitate state corporate crime; *glocality*, the materialisation of global trends in local spaces; *post-colonial interculturality*, unequal tensions between cultures; and *nature impairment* and *violence*, the intrinsic and unavoidable consequences of foreign extractivism. In sum, the chapter exposes the deep entrenchment of environmental degradation and victimisation in the global South in basic global social structures. To curb and prevent environmental harm, a radical change is necessary.

In Chapter 3, 'Mass Extraction and Green Crime Victimisation in Turkey', Halil Ibrahim Bahar sets out to test the links between international corporate intervention, national corruption and lack of democracy, and green victimisation. Bahar highlights the deleterious consequences of the extractive activities of transnational corporations in Southern locations, which he identifies as a continuation of centuries-long colonialist practices. The author characterises the ideology of Turkey's political elites as 'fetishizing growth with fatal consequences'. Indeed, the dominant political party in Turkey initiated a policy of deruralisation while paving the way for the activities of extractive transnational corporations.

Bahar's insights are based on two cases, the Soma Mine disaster and the Ikizdere stone pitch project. Both provide evidence of the collusion between corporations and political elites, and the dominant dynamics of internal colonialism. The chapter raises the question of how countries in the global South should organise politically to avoid Northern exploitation. Democratic projects that follow Western canons and adhere to the civilising project have been shown to fail at preventing neo-colonial environmental practices. So, how can we secure democracy under Southern scripts?

In Chapter 4, 'Environmental Exploitation and Violence Against Indigenous People in Mexico', Gabriela Gallegos Martínez, José Luis Carpio Domínguez, and Jesús Ignacio Castro Salazar study the main drivers of violent attacks on Indigenous people in Mexico and the repercussions of this for environmental conservation. The authors identify the main conflict leading to violence against Indigenous people as contested land. Indigenous people, in line with their environmental ontologies (Goyes, Abaibira et al., 2021), try to protect nature but in the face of corporations pursuing their economic interests. A third stakeholder, policymakers, address the conflict without solving it, as Chambliss (1979, 1989) predicted four decades ago. Violence against Indigenous people is thus engrained in the economic and legal structures of Mexico (as it is throughout Southern and Indigenous territories, regardless of their geographical position). At an even deeper level, the roots of Indigenous victimisation and genocide reside in the *coloniality* engraved in global social structures. In the hierarchy of lives and ways of life, Indigenous ones are near the bottom. All other structures, including the legal, are based on that premise.

Gallegos, Carpio, and Castro visibilise the above dynamics in their study. Inspired by the literature about the value of Indigenous cosmovisions for restorative relationships with nature, the authors provide a detailed analysis of the actors and dynamics of violence against Indigenous communities in Mexico. Using newspaper reports, the authors redress the lack of official statistics on murders of Indigenous environmental defenders. (The fact that no official statistics exist on the deaths highlight the lack of interest in protecting Indigenous groups and securing their rights.) They identify the legal operations of electricity production and mining as responsible for most of the violence against Indigenous populations. However, in an elegant return to the origins of Southern green criminology which three decades ago innovatively linked drug trafficking with harms against the environment (del Olmo, 1987; see also South in this collection), the authors document that drug cartels have in recent years become the main illegal group perpetrating direct violence against Indigenous people. Furthermore, the authors note the symbiotic cooperation between legal industries and illegal groups, which allows them to diversify to increase their profit (see also van Uhm & Nijman, 2022). Finally, this chapter highlights how the rights of Indigenous communities and threats to their wellbeing have been side-lined in academic literature and political circles in the global South.

In Chapter 5, 'Appropriating the Commons: Tea Estates and Conflict Over Water in Southern Malawi', Dave Mankhokwe Namusanya explores one of the clearest physical legacies of colonisation in Africa, tea estates, by probing into the coloniality of knowledge. Specifically, he analyses how relations between humans and water have been colonised for centuries, producing widespread socio-environmental harms. Namusanya documents the cultural clash between the North and the South: while for the former most of the elements in nature are resources to exploit, for the latter they are complex socio-environmental phenomena with meanings beyond their physicality (see also Goyes & Sollund, 2018). Upon the arrival of colonisers in Africa, the animist view of nature was disrupted. As in all the chapters in this volume, Namusanya's study provides evidence of the collusion between government and private enterprise to advance their economic interests regardless of the ecological and social harms of their actions. This chapter is a clear example of epistemological violence and its role in environmental harm.

In Chapter 6, 'Political Economy and the Government Attack on Sharks: A Non-Speciesist Southern Green Criminology', Reece Walters and Amy Couper examine the colonial violence done to non-human animals in a study of human relations with sharks. They highlight the Australian Indigenous view of sharks as intrinsically valuable because of their magnificence, power, and place in the ecosystem. In contrast, in the political economy of the North, non-human animals are resources for trade, leisure, and consumption. As with other 'resources', Northern ideology dichotomises life by separating nature from humanity and the developed from the Indigenous. Colonisers assume that the Indigenous connection with non-human animals as part of nature makes them inferior (see Goyes, Abaibira et al., 2021). Using Western Australia as a case study, the authors demonstrate that governments marginalise Indigenous knowledge even though it better preserves nature. The authors highlight the value of Indigenous environmental ontologies for the protection of non-human species particularly because of the harms produced by Northern views of non-human animals. In sum, this chapter contributes to the establishment of a non-speciesist Southern green criminology (see also Goyes, 2015) and demonstrates that coloniality also informs human relations with non-human animals.

Responses to Environmental Crime in the Global South

In Chapter 7, 'Green Potential in the Global South: The *Phulbari Movement* in Neo-Liberal Bangladesh', Nikhil Deb and Avijit Chakrabarty Ayon study the collective opposition of Bangladeshi people to extraction coal mining in Phulbari, which threatened to displace the Indigenous population and pollute the environment. This study also shows how the transnational Asia Energy Corporation colluded with the Bangladeshi government to advance the project by employing the discourses of development and economic growth (both part of the modern Western script) to persuade locals to support the mine. Furthermore, Asia Energy Corporation used the familiar tactics of manipulation and intimidation, widely used in the global South (Menton & Billon, 2021; Weinstock, 2017): bribing leaders to divide the community, criminalising social protest, and using physical violence.

However, the focus of the study is on the politics of resistance. The authors show that land displacement, in which local inhabitants are relocated to less viable locations, leads to environmental injustice. Conversely, collective action against land-grabs contributes to environmental justice. Deb and Ayon's study is a prime example of *the imagination of the poor*—the many creative ways in which the economically impoverished not only practice environmentalism but also imagine ways of resistance not yet captured by sociological theorisation. They resist creatively out of a necessity to survive. This chapter is, therefore, an important reminder of how crucial networks of activism throughout the global South are to fight the power of Northern corporations and to document the systematic nature of environmental violence.

In Chapter 8, 'Latin American Green Criminology and the Limits of Restorative Justice: An Analysis of the Samarco Case', Cristina Rego de Oliveira, Daniela Arantes Prata, and Bruna dos Santos L. da Silva ask how environmental destruction and victimisation should be repaired in Southern countries. The authors use the case of the Fundão dam collapse in the Brazilian Doce River basin to illustrate the colonial roots of most major environmental catastrophes in the global South, as well as the difficulties in restoring what the victims—humans, non-humans, and ecosystems—have lost. The Samarco case is one of the most salient tragedies in the global South: the Fundão dam collapsed in 2015, and 40 million cubic metres of mining tailings travelled down the Doce River. A major difficulty in reparation programmes, the authors argue, is the dependence of Southern states on Northern ones and the consequential fragility of Southern institutions.

Rego de Oliveira, Arantes Prata, and dos Santos draw on the frameworks of restorative justice that suggest that a healthy way of solving conflicts includes the participation of the parties of the conflict in a non-hierarchical way, the sharing of the responsibility for the conflict, and the effective reparation of the harms, while considering the needs of victims, offenders, and the broader community. This restorative justice framework, however, has specific challenges when applied to environmental victimisation: the victims are multiple and varied, the causes are not easy to identify and isolate, and the magnitude of the consequences make them difficult to identify. Furthermore, the institutional weakness of states in the global South makes redress via formal judicial procedures more difficult to achieve for victims of environmental harm. And while restorative justice could be an alternative to formal civil judicial processes, its application in Brazil still follows the canons developed in the global North (another sign of pervasive coloniality). Furthermore, as Southern green criminologists have predicted, environmental destruction in Brazil is not only the consequence of direct neo-colonial practices but also an outcome of internal colonialism—the ecocidal and genocidal policies of local governments guided by a development agenda.

The deep question that this chapter raises is how can we redress the damages, considering that colonialism, neo-colonialism, and coloniality have shaped the dynamics of victimisation in the global South? The authors rightly indicate that since green crime in the global South is a consequence of long-term colonial structures, not only the immediate harm should be considered, but also the historical dynamics when making reparation. The dependency of the global South must, therefore, also be considered when implementing a restorative justice programme in instances of environmental victimisation.

In Chapter 9, 'Beyond Redistributive Justice: Listening to Environmental Victims' Demands in Brazil', Marília de Nardin Budó, Karine Agatha França, and Lorenzo Natali continue the debate about how to deal with conflicts and redress crimes involving corporations and local communities in the global South. The authors warn about the undesirability of using the state legal system to address instances of environmental harm, considering that the state (in cooperation with corporations) is a central perpetrator. They also highlight the fact that in most environmental harms, victims remain invisible, are highly diverse (ranging from humans to non-humans), and have relatively little power. Perpetrators are corporations in collusion with governments and have economic, political, juridical, and epistemological power. Usually, the discourse and ideology of development is used as a category to neutralise or deny responsibility.

For a solution to these challenges, the authors build on the principle of Southern green criminology that decolonial action should be based on the voices of those suffering colonial, environmental victimisation. Using environmental disasters in Brazil, they develop a proposal for the retribution and reparation of green crimes. De Nardin Budó, Agatha, and Natali find that environmental victims commonly feel abandoned, neglected, and unacknowledged, do not trust state institutions, feel they will never be adequately compensated for the harms they suffer, and want reparation rather than vengeance (see also Franko & Goyes, 2023). Reparation, explain the authors, has to do with simultaneously looking backward and forward, building memory and acknowledging the harms, and making the victims feel seen and heard. The chapter makes visible the dimensions of environmental victimisation and the need for a restorative programme that equals such magnitude—a framework that goes beyond standard judicial lines and approaches peace processes in war settings.

In Chapter 10, 'Pop Culture as Environmental Education in Japan: The Case of Hayao Miayzak's *Kaze-no-tani-no-Naushika*', Orika Komatsubara explores non-orthodox forms of environmental education as a way to heighten solidarity with victims of environmental crime and prevent future catastrophes. Komatsubara asserts that popular culture, in its creativity, can capture the views of marginalised populations that have been environmentally victimised. Pop culture, such as Japanese manga, conveys ideas and worldviews that are otherwise difficult to explain to the wider audience. The wide reach of manga makes it a powerful tool to increase awareness about environmental destruction and increase empathy for its victims. This chapter is thus a celebration of the creativity of the marginalised and their usefulness in the search for alternative ways of interacting with nature (see also Escobar, 2007).

Global Dialogues About Crime and Destruction in the South

The question of how to relate to knowledge production in other parts of the world appears frequently in decolonial and southern theorisation. Some propose discounting the knowledge coming from the North altogether. Yet that contradicts the spirit of decolonial action. Santos (2014) proposed 'cultural translation' as a decolonial tool for creating non-hierarchical communication between different epistomologies. Santo's cultural translation recognises the communication barriers between knowledge traditions, which often result in a situation of no-comunication. This is an undesirable outcome, according to Santos, given that every knowledge is incomplete and can benefit from the epistemologies of others, particularly when gathering tools to fight against (colonial) impositions (see also Goyes, 2018a). Southern green criminology seeks to surmount this challenge by exchanging knowledge and insights with other epistemological traditions in a non-hierarchical way. The two final chapters in the collection tease out that practice.

In Chapter 11, 'Revisiting Rosa: Eco-Bio-Genocide, Drug Wars, and Southern Green Criminology', Nigel South documents how his proposition for a green criminology was taken up in the innovative work of Latin American criminologist Rosa del Olmo (1998), who analysed the war on drugs in terms of eco-bio-genocide. She concluded that the actions of the USA to eradicate drug crops were a crime committed against Latin America, its peoples, and ecosystems. With her contribution, del Olmo anticipated many of the debates later undertaken by green criminologists in the North. Simultaneously, Nigel South's interest in drug issues and his cosmopolitan imagination leads him to inquire about the harms surrounding drugs worldwide. South's research agenda, significantly inspired by del Olmo, connects the traditional field of drug research in criminology with broader concerns about environmental destruction and victimisation.

This chapter reveals the crucial contributions that the voices from the geographical and metaphorical South make to criminology. People in the South lead unique lives because of their geopolitical position, making their view of society different from that of the North's. Remembering the Northern-centric criminology of the 1980s, Nigel South (one of the most influential criminologists worldwide) laments the Northern bias in criminology and wonders about the consequential knowledge losses: 'one obvious if uncomfortable question to ask is whether the concept [of eco-bio-genocide] would have been more widely adopted if proposed from outside the borders of Latin American criminology?' He also problematises the appropriation by Northern scholars of the Southernising project through inclusive-sounding, critical concepts that once again give Europe and North America the lead on how the South should think and feel—not very decolonial. In sum, Nigel South takes issue with the invisibilisation of Southern contributions to criminology due to the biases present in 'international' criminology. He concludes, however, with a note of hope regarding the ongoing conversations that acknowledge Northern shortcomings and Southern contributions and potential.

In Chapter 12, 'Colonialism, Knowledge, and the White Man's Burden', Rob White continues the debate about the role of the 'white man' in processes of decolonisation. In the chapter title, he alludes to Kipling's idea of Western superiority and his support of colonialism: colonisers, making use of their assumed superior capacities, will rescue the rest of the world. Rob White acknowledges that centuries of oppression and exploitation have given the North the privileged position that it has today, including in the creation of knowledge, so he asks a question regarding what the offspring of the colonisers should do to repay their debt. In a rhetorical manoeuvre, the author turns the 'white man's burden' on its head, making it the white man's responsibility to respond to and repair colonial harms.

Rob White begins with foundational reflections. While a deep connection with nature has been attributed to Indigenous peoples as a definitive characteristic, it does not universally apply to them, nor is it not exclusive to them. The *identity* of a person—including their *worldview*—is not static; it depends on social structure, situation, and agency. So, a first step towards decolonising is altering the structural conditions to enable greater connection to the natural world. Implicitly alluding to the concept of coloniality, Rob White also recognises that *knowledge* plays a significant role in sustaining colonial structures and that Northern knowledge is usually positioned as valid while other forms of knowledge are exotic, circumstantial, or simply not-knowledge. Indigenous knowledge is, when seen, appropriated and converted to serve Northern interests. The author notes that much of the current decolonial action has been filled with symbolism and exhortation, but authentic decolonial action has to do with a worldview, identity, and knowledge that embrace decolonial action, learning from others, and transforming the self. He also remarks that scholarly decolonising action deals with undoing the dominant structures of knowledge production, which currently follow a neo-liberal script that disciplines scholars into a way of being and thinking and therefore makes it difficult for real decolonial action.

The Future of Southern Green Criminology

Argentinean criminologist Raúl Zaffaroni (2022) writes about the cultural criminal patrimony of humanity:

> The experience of atrocious mass crimes has been denied or neutralised but it has not disappeared; humanity sees the more or less repressed memories of what is best not to think about. However, those crimes accumulate slowly and silently and become a significant cultural criminal patrimony of humanity … However, the explosion [of the voices of the victims] does not stop the atrocious crimes and the process of denying them or normalising them continues through the use of ideological and technological tools, thus keeping alive the cycle of cultural criminal patrimony.

One significant component of that patrimony is the *environmental criminal inheritance of humanity*, which is necessary to see, acknowledge, document, and understand if its cycle of violence is to be curbed. Southern green criminologists are tasked with not allowing the world to deny, neutralise, or normalise colonially driven environmental destruction and victimisation. To continue along this promising path, this criminology must pay attention to three issues that this volume does not address in detail, if at all:

1. 'Blue', or marine, matters;
2. Pervasive gender inequalities in environmental victimisation and in criminological production;
3. The narrow scope of voices producing criminological knowledge.

We, as Southern green criminologists, must sustain our indignation and expand our critical thinking (Freire, 2001, 2017/1969) not to export to the global North the conflicts the *Environmental Justice Atlas* has identified in the global South, but to lead the world along a path of less destruction, less conflict, and less victimisation—our Southern voices must be heard.

References

Aas, K. F. (2013). *Globalization & crime* (2nd ed.). Sage.

Amin, S. (1997). *Capitalism in the age of globalization: The management of contemporary society*. Zed Books.

Amnesty International. (2016). *Trafigura: A toxic journey*. Amnesty Interntional. Retrieved October 30, from https://www.amnesty.org/en/latest/news/2016/04/trafigura-a-toxic-journey/#:~:text=The%20toxic%20waste%20was%20created,petroleum%20product%20called%20coker%20naphtha

Arroyo-Quiroz, I., & Wyatt, T. (Eds.). (2018). *Criminología verde en México*. CRIM-UNAM.

Bjørklund, I. (2020). The Áltá action as a Sámi conflict: Forty years later. In K. García-Antón, H. Gaski, & G. Guttorm (Eds.), *Let the river flow: An Indigenous uprising and its legacy in art, ecology and politics* (pp. 39–50). OCA.

Böhm, M. L. (2018). *The crime of maldevelopment: Economic deregulation and violence in the global South*. Routledge.

Böhm, M. L. (2020a). Criminal business relationships between commodity regions and industrialized countries: The hard road from raw material to new technology. *Journal of White Collar and Corporate Crime, 1*(1), 34–49. https://doi.org/10.1177/2631309x19887681

Böhm, M. L. (Ed.). (2020b). *Empresas transnacionales, recursos naturales y conflicto en América Latina: Para una visibilización de la violencia invisible [Transnational corporations, natural resources and conflict in Latin America: To make visible the invisible violence]*. Universidad de Buenos Aires.

Briggs, D. (2023). Breaking the shackles of academic capitalism: Academic life, liberation and ethnographic innovation. In R. Faria & M. Dodge (Eds.), *Qualitative research in criminology: Cutting-edge methods* (pp. 19–36). Springer International Publishing. https://doi.org/10.1007/978-3-031-18401-7_2

Brisman, A., & South, N. (2013). Resource wealth, power, crime and conflict. In R. Walters, D. Westerhuis, & T. Wyatt (Eds.), *Emerging issues in green criminology: Exploring power, justice and environmental harm* (pp. 60–71). Palgrave.

Brisman, A., & South, N. (2014). *Green cultural criminology: Constructions of environmental harm, consumerism, and resistance to ecocide*. Routledge.

Brisman, A., & South, N. (2020). The growth of a field: A short history of a 'green' criminology. In A. Brisman & N. South (Eds.), *Routledge international handbook of green criminology* (2nd ed., pp. 39–51). Routledge.

Budó, M. D., Goyes, D. R., Natali, L., Sollund, R., & Brisman, A. (Eds.). (2022). *Introdução à criminologia verde: Perspectivas críticas, descoloniais e do Sul [Introduction to green criminology: Critical, decolonial and Southern perspectives]*. Tirant lo Blanch Brasil.

Bull, B. (2021). *Latin-Amerika i dag: Nye interesser og gamle bånd til USA, Kina, Russland, Midtøsten og Europa [Latin America today: New interests and old ties to USA, China, Russia, the Mid-East and Europe]*. Cappelen Damm Akademisk.

Carrington, K., Dixon, B., Fonseca, D., Goyes, D. R., Liu, J., & Zysman, D. (2019). Criminologies of the global South: Critical reflections. *Critical Criminology, 27*(1), 163–189.

Chambliss, W. (1979). On lawmaking. *British Journal of Law and Society, 6*(2), 149–171. https://doi.org/10.2307/1409765

Chambliss, W. (1989). State-organized crime—The American Society of Criminology, 1988 presidential address. *Criminology, 27*(2), 183–208. https://doi.org/10.1111/j.1745-9125.1989.tb01028.x

Coleman, C., & Moynihan, J. (1996). *Understanding crime data: Haunted by the dark figure*. Open University Press.

de Carvalho, S., Goyes, D. R., & Vegh Weis, V. (2020). Politics and Indigenous victimization: The case of Brazil. *British Journal of Criminology, 61*(1), 251–271. https://doi.org/10.1093/bjc/azaa060

Dean, C. E. (2017). Social inequality, scientific inequality, and the future of mental illness. *Philosophy, Ethics, and Humanities in Medicine, 12*(1), 10. https://doi.org/10.1186/s13010-017-0052-x

del Olmo, R. (1987). Aerobiology and the war on drugs: A transnational crime. *Crime and Social Justice, 30*, 28–44.

del Olmo, R. (1998). The ecological impact of illicit drug cultivation and crop eradication programs in Latin America. *Theoretical Criminology, 2*(2), 269–278.

Dimou, E. (2021). Decolonizing Southern criminology: What can the 'decolonial option' tell us about challenging the modern/colonial foundations of criminology? *Critical Criminology*. https://doi.org/10.1007/s10612-021-09579-9

Duckett, A. (2009). Trafigura story breaks. *TCE Today*. https://web.archive.org/web/20101213040553/http://www.tcetoday.com/trafigura

Escobar, A. (1995). *Encountering development: The making and unmaking of the Third World*. Princeton University Press.

Escobar, A. (2007). Worlds and knowledges otherwise. *Cultural Studies, 21*(2–3), 179–210.

Franko, K., & Goyes, D. R. (2019). Global ecological destruction. In K. Franko (Ed.), *Globalization & crime* (pp. 193–212). Sage.

Franko, K., & Goyes, D. R. (2023). *Victimhood, memory, and consumerism: Profiting from Pablo*. Oxford University Press.

Freire, P. (2001). *Pedagogia da indignação: Cartas pedagógicas e outros escritos [Pedagogy of Indignation: Pedagogigcal letters and other writings]*. UNESP.

Freire, P. (2017 [1969]). *La educación como práctica de libertad [Education, the practice of freedom]*. Siglo XXI.

Galaz, V., Crona, B., Dauriach, A., Jouffray, J.-B., Österblom, H., & Fichtner, J. (2018). Tax havens and global environmental degradation. *Nature Ecology & Evolution, 2*(9), 1352–1357. https://doi.org/10.1038/s41559-018-0497-3

Galaz, V., Crona, B., Österblom, H., Olsson, P., & Folke, C. (2012). Polycentric systems and interacting planetary boundaries—Emerging governance of climate change-ocean acidification-marine biodiversity. *Ecological Economics, 81*, 21–32. https://doi.org/10.1016/j.ecolecon.2011.11.012

Goyes, D. R. (2015). Denying the harms of animal abductions for biomedical research. In R. Sollund (Ed.), *Green harms and crimes: Critical criminology in a changing world* (pp. 170–188). Ashgate.

Goyes, D. R. (2016). Green activist criminology and the epistemologies of the South. *Critical Criminology, 24*(4), 503–518.

Goyes, D. R. (2017). Corporate lobbying and criminalization. *Crime, Law & Social Change, 69*, 401–419.

Goyes, D. R. (2018a). Green criminology as decolonial tool: A stereoscope of environmental harm. In K. Carrington, R. Hogg, J. Scott, & M. Sozzo (Eds.), *The Palgrave handbook of criminology and the global South* (pp. 323–346). Palgrave.

Goyes, D. R. (2018b). 'Tactics rebounding' in the Colombian defence of free seeds. *International Journal for Crime, Justice and Social Democracy, 7*(1), 91–107.

Goyes, D. R. (2019). *Southern green criminology: A science to end ecological discrimination.* Emerald.

Goyes, D. R. (2020). 'Little development, few economic opportunities and many difficulties': Climate change from a local perspective. *International Journal for Crime, Justice and Social Democracy, 9*(2), 170–182.

Goyes, D. R. (2021). Environmental crime in Latin America and Southern green criminology. In *Oxford encyclopedia of criminology and criminal justice.* Oxford University Press. https://doi.org/10.1093/acrefore/9780190264079.013.588

Goyes, D. R. (2022). Latin American green criminology. *Justice, Power and Resistance*, 1–18. https://doi.org/10.1332/cdqk2391

Goyes, D. R., Abaibira, M. A., Baicué, P., Cuchimba, A., Ñeñetofe, D. T. R., Sollund, R., South, N., & Wyatt, T. (2021). Southern green cultural criminology and environmental crime prevention: Representations of nature within four Colombian Indigenous communities. *Critical Criminology, 29*(3), 469–485. https://doi.org/10.1007/s10612-021-09582-0

Goyes, D. R., Komatsubara, O., Droz, L., & Wyatt, T. (2022). Green criminological dialogues: Voices from Asia. *International Journal for Crime, Justice and Social Democracy, 11*(1), 1–10.

Goyes, D. R., Mol, H., Brisman, A., & South, N. (Eds.). (2017). *Environmental crime in Latin America: The theft of nature and the poisoning of land.* Palgrave Macmillan.

Goyes, D. R., & Sollund, R. (2018). Animal abuse, biotechnology and species justice. *Theoretical Criminology, 22*(3), 363–383. https://doi.org/10.1177/1362480618787179

Goyes, D. R., Sollund, R., & South, N. (2019). Guest editors' introduction: Towards global green criminological dialogues: Voices from the Americas and Europe. *International Journal for Crime, Justice and Social Democracy, 8*(3), 1–5.

Goyes, D. R., & South, N. (2016). Land-grabs, bio-piracy and the inversion of justice in Colombia. *British Journal of Criminology, 56*(3), 558–577. https://doi.org/10.1093/bjc/azv082

Goyes, D. R., South, N., Abaibira, M. A., Baicué, P., Cuchimba, A., & Ñeñetofe, D. T. R. (2021). Genocide and ecocide in four Colombian Indigenous communities: The erosion of a way of life and memory. *British Journal of Criminology, 61*(4), 965–984. https://doi.org/10.1093/bjc/azaa109

Goyes, D. R., South, N., Sollund, R., & de Carvalho, S. (2021). Editors' introduction to the special issue, 'Southern criminologies: Methods, theories and Indigenous issues.' *Critical Criminology, 29*(3), 423–429. https://doi.org/10.1007/s10612-021-09586-w

Gutierrez, G. (1974). *A theology of liberation.* SCM Press.

Haug, S. (2021). *What or where is the 'global South'? A social science perspective.* LSE Impact Blog. Retrieved October 10, from https://blogs.lse.ac.uk/impactofsocialsciences/

Haug, S., Braveboy-Wagner, J., & Maihold, G. (2021). The 'global South' in the study of world politics: Examining a meta category. *Third World Quarterly, 42*(9), 1923–1944. https://doi.org/10.1080/01436597.2021.1948831

Hickel, J. (2020). Quantifying national responsibility for climate breakdown: An equality-based attribution approach for carbon dioxide emissions in excess of the planetary boundary. *The Lancet Planetary Health, 4*(9), e399–e404. https://doi.org/10.1016/S2542-5196(20)30196-0

Hornborg, A. (2009). Zero-sum world: Challenges in conceptualizing environmental load displacement and ecologically unequal exchange in the world-system. *International Journal of Comparative Sociology, 50*(3–4), 237–262. https://doi.org/10.1177/0020715209105141

Is science only for the rich? (2016). *Nature, 537*(7621), 466–470. https://doi.org/10.1038/537466a

Johannessen, L., Rafoss, T. W., & Rasmussen, E. B. (2020). *Hvordan bruke teori? Nyttige verktøy i kvalitativ analyse [How to use theory? Useful tools in a qualitative analysis]*. Universitetsforlaget.

Kumar, K. (2021). Colony and empire, colonialism and imperialism: A meaningful distinction? *Comparative Studies in Society and History, 63*(2), 280–309. https://doi.org/10.1017/S0010417521000050

Lessenich, S. (2019). *Living well at others' expense: The hidden costs of Western prosperity*. Polity.

MacManus, T. (2016). The denial industry: Public relations, 'crisis management' and corporate crime. *International Journal of Human Rights, 20*(6), 785–797.

Mahoney, J. (2012). *Colonialism and postcolonial development: Spanish America in comparative perspective*. Cambridge University Press.

Martinez-Alier, J. (2018). Ecological distribution conflicts and the vocabulary of environmental justice. In V. Dayal, A. Duraiappah, & N. Nawn (Eds.), *Ecology, economy and society: Essays in honour of Kanchan Chopra* (pp. 187–204). Springer Singapore. https://doi.org/10.1007/978-981-10-5675-8_11

Melchiors, R. B. (2022). *Criminologia verde: Grandes corporações e danos socioambientais no Sul global [Green criminology: Big corporations and socioenvironmental harms in the global South]*. Editora Blimunda.

Menton, M., & Billon, P. L. (Eds.). (2021). *Environmental defenders: Deadly struggles for life and territory*. Routledge.

Mol, H. (2017). *The politics of palm oil harm: A green criminological perspective*. Palgrave Macmillan.

Mondaca, E. (2017). The archipelago of Chiloé and the uncertain contours of its future: Coloniality, new extractivism and political-social re-vindication of existence. In D. R. Goyes, H. Mol, A. Brisman, & N. South (Eds.), *Environmental crime in Latin America: The theft of nature and the poisoning of the land* (pp. 31–56). Palgrave Macmillan.

Moosavi, L. (2018). A friendly critique of 'Asian criminology' and 'Southern criminology.' *British Journal of Criminology, 59*(2), 257–275. https://doi.org/10.1093/bjc/azy045

Quijano, A. (2000). Coloniality of power, Eurocentrism, and Latin America. *Nepantla: Views from South, 1*(3), 533–580.

Quijano, A. (2007). Coloniality and modernity/rationality. *Cultural Studies, 21*(2–3), 168–178.

Ramírez, R. (2021). Pacific Islanders have been fighting environmental crises for centuries, if only the world would notice. *Grist*. https://grist.org/fix/oceans/pacific-islands-climate-change-innovations/

Readfearn, G. (2022, 14 July). CSIRO joins deep-sea mining project in Pacific as islands call for industry halt. *The Guardian*. https://www.theguardian.com/australia-news/2022/jul/14/csiro-joins-deep-sea-mining-project-in-pacific-as-islands-call-for-industry-halt

Reardon, K., Padfield, R., & Salim, H. K. (2019). 'Consumers don't see tigers dying in palm oil plantations': A cross-cultural comparative study of UK, Malaysian and Singaporean consumer views of palm oil. *Asian Geographer, 36*(2), 117–141. https://doi.org/10.1080/10225706.2019.1621187

Reuters. (2022). Mexico's Pemex increases crude exports to U.S., cuts shipments elsewhere. *Reuters*. https://www.reuters.com/markets/commodities/mexicos-pemex-increases-crude-exports-us-cuts-shipments-elsewhere-2022-06-27/

Ruggiero, V., & South, N. (2013). Toxic state-corporate crimes, neo-liberalism and green criminology: The hazards and legacies of the oil, chemical and mineral industries. *International Journal for Crime, Justice and Social Democracy, 2*(2), 12–26. https://doi.org/10.5204/ijcjsd.v2i2.115

Rust, S. (2019). How the U.S. betrayed the Marshall Islands, kindling the next nuclear disaster. *LA Times*. https://www.latimes.com/projects/marshall-islands-nuclear-testing-sea-level-rise/

Santos, B. d. S. (2014). *Epistemologies of the South: Justice against epistemicide*. Paradigm Publishers.

Smithers, R. J., Blicharska, M., & Gutiérrez, J. M. (2016). Inequality: Span the global divide. *Nature, 539*(7627), 31–31. https://doi.org/10.1038/539031c

Snipes, J. B., Bernard, T. J., & Gerould, A. L. (2019). *Vold's theoretical criminology*. Oxford University Press.

Statista. (2022). *Production volume of palm oil worldwide from 2012/13 to 2021/22*. Retrieved December 11, from https://www.statista.com/statistics/613471/palm-oil-production-volume-worldwide/

Stretesky, P. B., Long, M. A., & Lynch, M. J. (2014). *The treadmill of crime: Political economy and green criminology*. Routledge.

Temper, L., del Bene, D., & Martinez-Alier, J. (2015). Mapping the frontiers and front lines of global environmental justice: The EJAtlas. *Journal of Political Ecology, 22*, 255–278. https://journals.librarypublishing.arizona.edu/jpe/article/id/1932/

van Uhm, D. P., & Nijman, R. C. C. (2022). The convergence of environmental crime with other serious crimes: Subtypes within the environmental crime continuum. *European Journal of Criminology, 19*(4), 542–561. https://doi.org/10.1177/1477370820904585

Varpio, L., Paradis, E., Uijtdehaage, S., & Young, M. (2020). The distinctions between theory, theoretical framework, and conceptual framework. *Academic Medicine, 95*(7), 989–994.

Wallerstein, I. (2004). *World-systems analysis: An introduction*. Duke University Press.

Weinstock, A. M. (2017). A decade of social and environmental mobilization against mega-mining in Chubut, Argentinian Patagonia. In D. R. Goyes, H.

Mol, A. Brisman, & N. South (Eds.), *Environmental crime in Latin America: The theft of nature and the poisoning of the land* (pp. 141–162). Palgrave.

Yeager, M. G., & Smith, J. L. (2017). Global pollution, multinational oil companies and state power: The case of Yaiguaje v. Chevron Corporation. In D. R. Goyes, H. Mol, A. Brisman, & N. South (Eds.), *Environmental crime in Latin America: The theft of nature and the poisoning of the land* (pp. 117–140). Palgrave.

Zaffaroni, E. R. (2022). *Colonialismo y Derechos Humanos: Apuntes para una historia criminal del mundo [Colonialism and human rights: Notes for a criminal history of the world]*. Penguin.

Part I

Drivers of Green Crime in the Global South

2

The State-Corporate Crime of Extractive Industries

María Laura Böhm

Introduction

Many years ago I felt for the first time the need to learn more about the links between the economic interests of the global North (a term whose use was not so widespread at the time) and the direct involvement with extractive activities in the global South. From the well-known blood diamonds obtained in African conflict contexts and sold in expensive European jewellery stores, to the ancient tradition of rubber extraction in South America for its industrial use in other contexts, I gradually began to learn about these cases and study them avidly. After visiting and talking with members of Indigenous communities, the need for this study became more evident. Concern and curiosity became criminological research, and some of its results are the subject of the present text.

M. L. Böhm (✉)
Ludwig-Maximilians-Universität München, Munich, Germany
e-mail: marialauraboehm@jura.uni-muenchen.de

D. R. Goyes (ed.), *Green Crime in the Global South*, Palgrave Studies in Green Criminology, https://doi.org/10.1007/978-3-031-27754-2_2

This chapter presents the findings of an extended study on the criminal traits of resource extraction projects in four Latin American cases: ChevronTexaco in Ecuador, Mariana in Brazil, Valle de Siria in Honduras, and Lote 8 in Guatemala.[1] For the analysis, the concept of *state-corporate crime* (scc) is used, which refers to a form of crime that results from the interaction of the interests of political and economic actors. Such an interaction leads to behaviour that is not always criminalised, but which nevertheless causes serious social and environmental damage. In this chapter, the concept scc is specifically applied to industries extracting raw materials. Because of this, the phenomenon and analytical tool are presented as *raw materials-related state-corporate crime* (rmr-scc).

Following this introduction there is a brief overview of commodity industries, a short conceptualisation of scc, and a summary presentation of four extractive cases. The analysis is then carried out in the following two sections, which focus on (1) the systematic presentation of the information obtained on activities, interactions, and damage in the said cases, and on (2) the relevant characteristics of rmr-scc. The aim of this chapter is to contribute to the study and development of the concept of scc from a Southern perspective.

Extractive Industries in the Twenty-First Century

The extractive industries dedicated to raw materials are those branches of industry that have the exploration, mining, processing, and transport of certain commodities as their main activity (Hillebrand, 2016, p. 12 ff.).

The extractive resource industry in the twenty-first century has characteristics that this industry did not have 500 or even 80 years ago (Peluso & Lund, 2011, p. 669). Raw materials have increased in demand

[1] The project 'Transnationale Wirtschaftskriminalität als globales *State-Corporate Crime*. Rohstoffbeziehungen mit Lateinamerika auf dem kriminologischen Prüfstand' is the doctoral research submitted by me at the Law Faculty (Chair for Criminal Law and Criminology) of the Ludwig-Maximilians Universität München and which was rated as *summa cum laude* in May 2022 and was published by Nomos Verlag in that year.

(Bebbington, 2012, p. 8; Rachman, 2016, p. 4), and the mining methods and the regions in which the raw materials are prospected and mined have changed in some cases (Svampa & Viale, 2014, pp. 171 ff., 299 ff.). Latin America, for example, was just as popular as an extractive region 500 years ago as it is today: back then it was exploited for its gold, silver, and rubber (Galeano, 1984, p. 15 ff., pp. 139–144; IDAMHO 2012, p. 11 ff.), while today it is exploited for lithium for electric cars and batteries (see Gómez Lende, 2017, p. 160 ff.) and for African palm for biodiesel (see Dammert, 2019, p. 53 f.). Increased demand and new corporations come from the Americas and Asia (Palacios et al., 2018, p. 12; Rachman, 2016, pp. 4 ff.; Svampa & Viale, 2014, p. 359), and not just Europe.

New projects in the raw materials industry often bring visible advantages, not only for the companies, but also for the population. However, the negative consequences can be much more than the positive ones. Although some methods today are gentler and more respectful of workers, it can be argued that the sheer volume, numbers, and aggressive acceleration of a large number of today's mining projects have an extremely negative impact beyond the project installation. As will be seen, the local population, neighbouring places, flora, fauna, land, water, and air are affected; social, economic, and even institutional damage is caused as well; forms of eviction of residents, the involuntary resettlement of a community through fraudulent promises, the abolition of labour rights, and the funding of armed groups involved in committing serious crimes are common too (Böhm, 2016, 2019; Goyes & South, 2016a, 2016b, 2019).

These harmful activities can be acts of commission or omission, and they can be legal or illegal (Eisenberg & Kölbel, 2017, §47, marginal note 4; Fernández Steinko, 2008, 2013; Huisman, 2008; Huisman & van Sliedregt, 2010). Just because we are talking about harmful activities does not mean these activities are illegal (Godson, 2003, Huisman, 2008, Huisman & van Sliedregt, 2010, p. 826; Ruggiero, 1996). They can be perfectly legal and still have serious harmful consequences, such as using certain problematic chemicals to extract gold (Bundschuh et al., 2012, pp. 18, 23; Pieth, 2019, p. 126), or fencing a plant and its land, which then prevents the daily access of the local population to its water resources (Opel, 2016, pp. 20, 22 s., 47; Sandoval Vinelli, 2018).

The economic actors that conduct these activities are often transnational. They often perceive and take advantage of the weaknesses of resource regions, such as Latin America, in terms of economy, political institutions, legal order, and society organization, and consequently take over the regions as a profitable economic area for business (Bruckmann, 2017; Bussmann, 2010, p. 20).

States often do not prevent these activities and their serious consequences. This is a too general statement, but it describes the idea of 'regulatory consideration (or negligence)' (Eisenberg & Kölbel, 2017, §47, marginal note 18) by the state and reflects the reality of many source regions of the global South (Bebbington, 2012, p. 6; Ezeonu, 2015, p. 99; Svampa & Viale, 2014, pp. 217, 360). The executive branch promotes certain extractive industries, supports them with taxes—often to the detriment of local people—and offers other countries privileged conditions without imposing administrative sanctions on violators (Bruckmann, 2017; Gorenstein, 2016, p. 20 ff.; Gudynas, 2015, p. 22; Opel, 2016, p. 187; Raskovsky, 2020, p. 156 ff.; Zabalo, 2008). The state and its activity or negligence shows itself like a fractal phenomenon in the most diverse forms, repeating a reluctance at its different levels and realms. Under-criminalisation (Barak, 2017, p. 44 f.; Friedrichs, 2006, p. 128; Henry, 1991) is a legally sly version of this diffuse form of non-control (Eisenberg & Kölbel, 2017, §47, fn. 239). In this context, formal legal categories are inappropriate for the study of the problem. This is central to the application of the concept of scc, which moves in the grey areas between legality and illegality. Normative and factual impunity are intertwined.

The interaction between state and private actors partly explains a sometimes apathetic and sometimes promoting attitude by states. The phenomenon of the so-called 'revolving doors', for example, explains the mutual subterfuge between private and state actors (Barak, 2016, p. 139; Hoogenboom, 2010, p. 178 f.; Kölbel, 2019, p. 265; Michalowski & Kramer, 2006, p. 2; Zaitch & Gutiérrez Gómez, 2015, p. 392 f.). But it also explains the detailed knowledge of how government works that private companies often have, and it facilitates the development of policy influence in the legislative arena and strategies for avoiding detection and sanctions.

State-Corporate Crime

The concept of scc was first introduced by Ronald Kramer in 1990 (Kramer, 1992, 2006). The problem he addresses is the socially damaging intersection of corporate interests and state actors. Before Kramer coined the concept, he had already researched the topic, along with Raymond Michalowski years earlier. State-corporate crime is about 'the interaction between governmental agencies and private businesses' (Matthews & Kauzlarich, 2000, p. 282).

The definition given by Judy R. Aulette and Raymond Michalowski, who revised previous proposals, is the one adopted here and used for its application to commodity relationships:

> State-corporate crimes are illegal or socially injurious actions that result from a mutually reinforcing interaction between (1) policies and/or practices in pursuit of the goals of one or more institutions of political governance and (2) policies and/or practices in pursuit of the goals of one or more institutions of economic production and distribution. (Aulette & Michalowski, 1993, p. 175; 2006, p. 47)

A fundamental contribution of this approach is that it suggests that when investigating crime at the company level, it must be borne in mind that large companies and their economies cannot grow and their actual networks of activities cannot develop unless they are directly linked to the activities and to the interests of the state (Kramer et al., 2002, p. 266).

After decades of fundamental development in Anglo-Saxon countries, the concept and its use entered an internationalisation phase in the 2010s. On the one hand, authors from different regions were involved in theorising the connection between corporations and states, and, on the other hand, the questions raised in the studies increasingly pointed to the need to work on the interactions between global actors. The new approach soon zeroed in on the environmental problems and environmental damage caused by transnational scc and, in doing so, it further developed the harm perspective. The affected populations were examined more closely under the lenses of human rights violations and the unequal processes and effects of scc from a (de)colonial perspective from Africa, Asia, Oceania, and Latin America (Böhm, 2019; Bran-Guzmán, 2017; Brisman & South, 2018; Brisman et al., 2017; Ezeonu, 2015; Friedrichs & Rothe, 2015; Goyes et al., 2017, 2019, 2022; Middeldorp, 2016b; Torres Wong, 2018; Zaitch & Gutiérrez Gómez, 2015).

The focus of the work in this new period is environmental damage and social damage—including the denial of fundamental rights (Cunneen, 2015; Cunneen & Tauri, 2017, pp. 433 ff.; Stańczak, 2017, p. 235; Torres Wong, 2018).

Currently, the concept of scc seems to be on the verge of a specialisation phase, in which more and more specific types of crime and areas are examined. In the case of the raw materials industry and raw materials relations, several works have dealt with the subject (Ezeonu, 2015, 2018; Torres Wong, 2018; Zaitch & Gutiérrez Gómez, 2015). However, it now seems important to study scc in a systematic way—to subject the field to a systematic analysis of its specific criteria. This is the aim of the research briefly presented in this chapter: the application of the concept of scc to the study of transnational resource extraction relations.

Four Cases of Raw-Materials-Related State-Corporate Crime

Four cases are briefly presented in this section to show the diversity of situations and actors which can be involved. They were all analysed in detail in the research project on which this chapter is based.[2] These cases are not explained in detail here but are only summarised as situational and real-world information in order to render more comprehensible the

[2] From the perspective of invisible violence and to explain the criminological proposal of 'crime of maldevelopment', the cases of Chevron/Texaco, Valle de Siria, and Lote 8 have already been—along with five other cases—deeply analysed in Böhm (2019).

conceptual parts offered in Sections "The Criminological Systematisation of Incidents, Interactions, and Damages" and "Characteristics of Raw-Materials-Related State-Corporate Crime".

ChevronTexaco in Ecuador: Between 1964 and 1992, Texaco, Inc. (later ChevronTexaco), in cooperation with the Ecuadorian state oil company CEPE (later Petroecuador), explored, constructed, and exploited pipelines between Lago Agrio and the port of Esmeraldas in the Sucumbíos province of Ecuador. Oil extraction was done with low-quality technology, although at the time Texaco had the most advanced technology and was using it in the United States. The company's chosen methods in Ecuador meant using imperfect structures for transporting and processing crude oil and storing waste, leaving the water and soil of the jungle infiltrated with toxic substances. The consequences of the oil production project can be summarised as: serious environmental damage, degenerative diseases with a particular incidence of cancer, and economic, social, and cultural impairment of the living conditions of the local population.[3]

Mariana in Brazil: On 5 November 2015, the Dam Fundão of the Samarco Mineração S.A. (Joint venture of Brazilian and Anglo-Australian capital) mining landfill collapsed. Between 50 and 60 million cubic metres of toxic waste flooded the valley of the Santarém creek (Minas Gerais). The toxic mud wave inundated the Rio Gualaxo do Norte river basin and completely destroyed the village of Bento Rodrigues. After reaching the Rio do Carmo River, the mudslide took three hours to reach the Rio Doce River, from where it then took another two weeks to travel 680 km down to the Atlantic Ocean, where it dispersed. Improper management and supervision of the mining landfill resulted in the deaths of 19 people, the health impairment of entire communities, serious environmental damage and negative impacts on flora and fauna, as well as social, cultural, and economic losses, and last but not least, the complete wiping out of an entire community.[4]

[3] The main sources for this compact summary are Kimerling (1991, 1994, 2006), Donziger et al. (2010), and Ortiz (2011). For more detailed information and sources see Böhm (2019).

[4] The main sources for this compact summary are Russau (2016), Galindo de Fonseca and Galindo de Fonseca (2016), Brasil and Pires (2017), and Da Silva et al. (2017). For more information and detailed sources see Böhm (2019).

Valle de Siria in Honduras: Valle de Siria (Francisco Morazán, Honduras) is located in an exclusively agricultural area known both for its scarcity of surface water and for the extreme poverty of its inhabitants. In 1995, the new gold rush exploded here, attracting attention from international markets and increasing their interest in the region. In this context, exploration work began in 1995 and then open pit mining for gold deposits started between 2000 and 2007 by Mar West Resources, then Goldcorp (Canada) through its Honduran subsidiary Entre Mares. Mining activities ceased in 2008 without a proper mine deactivation programme. The damage caused by the project includes the displacement of communities, water and soil pollution, damage to flora and fauna, and various diseases, some of which are deadly.[5]

Lote 8 in Guatemala: In 2006 and 2007, the Compañía Guatemalteca de Níquel (CGN) forced the relocation of Q'eqchi' communities from the communes of Panzós (Alta Verapaz district) and El Estor (Izabal district) to the Polochic Valley to carry out mining activities. In this context, at least 11 women from the Q'eqchi' community of Lote 8 (El Estor) were raped by security forces and privately hired men. The women only filed the rape complaints after the killing of Indigenous teacher and community leader Adolfo Ich Chaman. This assassination also occurred in the context of the company's expansionist manoeuvres in the region. CGN was then a subsidiary of transnational mining company Skye Resources, which was purchased by HudBay Minerals in 2008, taking legal responsibility of Skye, which was headquartered in Toronto.[6]

[5] The main sources for this compact summary are Middeldorp (2016a, 2016b), IDAMHO (2012), Trucchi (2014), Torres Funes (2016), and Cálix (2017). For more detailed information and sources see Böhm (2019).

[6] The main sources for this compact summary are Méndez Gutiérrez (2013), Ramírez Parra (2014), and Alonso (2020). For details on the company merger, see Crystal et al. (2014, p. 6 ss.). For more detailed information and sources see Böhm (2019).

The Criminological Systematisation of Incidents, Interactions, and Damages

Events

For the analysis of the cases and their occurrences, the collection of data was guided by the following questions: Where and when did these events unfold? Who was involved? How deviant were the acts? From this search, information has been classified according to the place and time of the event, the function or position of power of the perpetrators or the actors involved, and the intensity of the deviance. According to each criterion, various possibilities have emerged. The results are summarised in Table 2.1.

The criminologically relevant events related to raw materials extraction took place very specifically in the place of extraction (1) and also in a somewhat broader area if they occurred in the vicinity of the mining or oil production (2). However, the actions did not only take place amid nature or near the facilities, but also in urban closed spaces of the state authorities and companies when decisions were made (3). Events also occurred in other zones in the source country, even if they were not directly undertaken by economic and/or political actors—but by contracted security forces, for example (4). In company or official rooms or meetings in industrialised countries,[7] that is, where the parent company was located, key decisions were also often made by the company or the politicians, which are to be regarded as part of the events (5).

If one considers the criterion of time of occurrence, both before (6), during (7), and after (8) the extraction of the raw materials in question, actions were taken and omissions took place that are also to be understood as part of the overall criminological relevance of the events. There was always a preparation phase, then the actions themselves and later

[7] The term 'industrialised country' refers to the countries involved in each case in the raw material importation for its use in industrial activities and technology development. In this sense, China, India, and Brazil are in many cases industrialised countries in interaction with source countries.

Table 2.1 Analytical systematisation of events using the example of the *cases ChevronTexaco, Mariana, Valle de Siria*, and *Lote 8*

A. Location of events	1. At the raw material extraction site 2. In the raw material extraction area 3. In company or official rooms/meetings in the source country 4. Elsewhere in the source country 5. In company or official rooms/meetings in industrialised country
B. Time of events	6. Before raw material extraction 7. During raw material extraction 8. After raw material extraction
C. Function/position of the perpetrators or the actors involved	9. Members of extractive companies in the source country 10. Members of extractive companies in industrialised countries 11. State actors from the source country 12. State actors from the industrialised country 13. Other actors from the source country 14. Other actors from the industrialised country
D. Deviance intensity	15. Slightly deviant 16. Medium deviant 17. Significantly deviant

the possible reactions, which were also consequences of the interactions carried out and which had longer-lasting effects.

Depending on the function or position of the perpetrators or the actors, it is interesting to note that a large number of actors were involved in the events. Members of the extractive companies in the source country were directly involved (9), somewhat more indirectly (or actually more invisible) was the participation of the members of the company who operated from the industrialised country and, for organisational reasons, were usually the main decision-makers (10). The state actors in the source country also showed direct involvement and enacted policies, decisions, and directives (11), while the state actors in the industrialised country participated more in the drafting of economic policies from a distance—although there were also cases of direct interference in the legislative processes of the source country (12). Other actors in both the source country (13) and the industrialised country (14) were also involved in the events and their actions were also important to the unfolding of the events.

According to the intensity of deviance, it was found that not all events were equally 'serious' regarding rights violations. Nevertheless, it has been shown that, taken together, the individual events had much power to cause damage. Apparently less deviant behaviours,[8] in the sense of purely formal discrepancies or mistakes (15), often led to acts of moderate deviance. Some of the latter were at the same time actions that should already have been regarded as infractions (16) and at the same time opened the way for the strongly deviant events that can be classified as simply 'criminal' (17).

The combinations of place, time, function, or position of the perpetrators and deviance of the actions were, of course, varied. These depended in large part on the types of interactions that advanced their interests.

Interaction of Interests

The cases under examination showed a mutually reinforcing interaction between policies and practices to pursue the goals of political institutions on the one hand and the policies and practices of institutions of economic production and distribution on the other. This link is the interaction already described in Aulette and Michalowski's concept of scc (Aulette & Michalowski, 1993). Interactions of this kind can be viewed from various perspectives.

Table 2.2 shows a systematisation of the findings.

Based on the political institutional interest on the part of the source country and the industrialised country, it can be seen that the source country in particular repeatedly considered the promotion and approval

[8] Here deviant behaviours are understood to be those that deviate from the social norm and, in some cases, are even behaviours that violate the law.

Table 2.2 Analytical systematisation of interactions of interests using the example of the *cases ChevronTexaco, Mariana, Valle de Siria,* and *Lote 8*

A. Political institutional interest on the part of the source country and the industrialised country	1. Necessity 2. Historical tradition 3. Opportunity for development
B. Economic institutional interest on the part of the extractive company and other companies	4. Necessity 5. Business cycle opportunism 6. Common economic behaviour in the extractive industry
C. Negotiation (im)balance in the interactions	7. Between the industrialised country and the source country 8. Between the source country and the extractive company 9. Between the source country and the local population 10. Between the extractive company and the local population 11. Between the local population and other social actors

of projects to be necessary (1). The interest of some actors has shown itself to be the continuity of a historical tradition of the raw materials and industrialised countries (2). In all cases, the projects were repeatedly justified with a development policy that was stated as a purpose by the governments in the source country and supported by the industrialised countries (3). However, it is questionable to what extent the need for development declared by the state actors could actually be promoted by these projects in the specific cases. Secondly, it is also questionable whether the extractive industry can be considered a motor of development at all. Beyond this question, however, it was undoubtedly observed in the cases examined that the promised development was not materialised—at least not thanks to these projects and much less to the benefit of the local population concerned.

The purpose of financial profit belongs to the nature of a business enterprise. In this sense, it can be seen in all cases that this general economic interest on the part of the companies ChevronTexaco, Samarco, Goldcorp, and HudbaySkye has driven their respective actions and activities.

In addition, specific interests and corresponding forms of communication are always present in all interactions which shaped the relationships.

Acting out of necessity is different from acting out of greed, and is different again from acting out of habit. In the overarching study, briefly presented in this chapter, the traits these interests show were traced. The forms of interaction vary depending on interest. According to the entrepreneurial institutional interest, two of these three specific interest modalities were identified in the study.

The category of necessity, which comes first in Table 2.2, was taken into account in data collection despite the improbability of its occurrence (4). It is possible that the companies examined initiated the corresponding project only out of necessity, or that the projects had to develop in such a damaging form only because there was no alternative. When collecting the data, the necessity was considered possible on the part of the company. Nonetheless, in none of the cases was there any indication that the projects had proceeded in the manner stated because this was so necessary or because the company had no other options available. Rather, it was two other specific categories of interests that were reflected in corporate participation and behaviour.

Economic opportunism was evident in all cases (5) and was always associated with extremely unbalanced interactions. The source countries had assured the companies very favourable economic conditions for their business and hardly any state control of the activities. This happened in the context of national economic crises, highest transnational demand, internal conflicts, or under unlawful governments, which meant for the companies the perspective of highly profitable and easy business—even by carrying out deviant activities. In the interactions, the companies were therefore often superior to their state partners at the negotiating table.

In addition to the specific opportunities that the economic, social, and institutional situation could offer, the extractive industry also shaped the interests and thus the forms of interaction on the part of companies due to its common economic behaviour (6). Large companies are usually subject to fewer controls than smaller ones. Their complex technical and organisational processes are hardly visible to third parties, and they are initially welcomed by the population, that is accepted without significant resistance. These circumstances were considered normal by the companies in their interactions. The rule was that socially acceptable practices that are common and followed by the company in their own country were considered by the same corporation as less important when operating in a foreign resource country.

Different levels of imbalance could be considered in the negotiations. The criterion of negotiation (im)balance does not exist in the original formulation of the concept of scc. However, data collection has shown a constant imbalance in the interactions, which is to be understood in terms of the *glocality* (see Section "Glocality") and the different positions of power of the subjects involved. Interactions between the countries of origin of the companies and their host countries are quite disparate and this can be traced back to the geopolitical power positions these states hold in the world economy and politics—Ecuador, Brazil, Honduras, and Guatemala do not belong to the global North's sphere of power as do the USA, Great Britain, Germany, Canada, or even Australia (7).

The interactions between authorities of the source country and the extractive companies were also unbalanced, and this was not least related to the nature of the interests (8). The driving force of corporate economic strength has even immediate repercussions in terms of changes in institutional structures of the source country.

The interactions between the state and its own population were even more unbalanced (9), since the population usually belonged to often forgotten Indigenous, rural, or poor populations and were not directly under the protective umbrella and supportive view of the state.

Due to this state neglect, it was to be expected that very strong discriminatory and unbalanced interactions would take place between the companies and the local population. With no government oversight and a relative lack of power, the local population was helplessly exposed to the companies. It was neither legally nor economically, and even less in cultural terms, on an equal footing that the discussions and negotiations with the companies took place (10).

In relation to the modalities of interaction of the affected population with other actors and social sectors, a double result is presented. On the one hand, there is clear empathy and support from social actors committed to collective causes, although they themselves are not directly affected (civil organizations, NGOs, universities). On the other hand, an attitude more identified with the interests and cultural guidelines transmitted by transnational companies is what explains the indifference of large sectors of the population towards the affected population (11).

The findings show that, in the cases analysed, these types of extractive projects and their implementation were made possible only by the interaction of interests between political and economic actors. These are typical in the sense of scc. On the one hand, these interactions of interests became concrete in the deviant events studied above and, on the other hand, caused serious social damage—as will be presented below.

Harm

The information collected and case data, reporting the negative impacts of extractive industries, were organised according to three criteria: by impact area; by impact magnitude; and by impact duration. Table 2.3 shows the proposed systematisation.

According to the criterion of area of impact, the damage is classified under the different areas of life or life domains that suffer from the serious consequences of corporate and political interactions with raw materials. The impact, however, is impossible to be clearly demarcated as its full magnitude may not be visible for many years. The first of these

Table 2.3 Analytical systematisation of harm using the example of the *cases ChevronTexaco, Mariana, Valle de Siria,* and *Lote 8*

A. Impact area	1. Environment and non-human life
	2. Life and human health
	3. Physical and psychological integrity
	4. Economy and society
	5. Culture and identity
B. Impact magnitude	6. Individual
	7. Intra-community
	8. Extra-community
	9. National
	10. Transnational
	11. Global
C. Impact duration	12. Short term
	13. Medium term
	14. Long term

impact areas to be considered under this criterion is the impact on the environment and non-human life (1). They come first because environmental damage brings about sickness and death to residents along with dismemberment and displacement of populations. The impact on life and human health is considered second place, whereby not only death but also long-term disease processes are included (2); further, the impact on physical and psychological integrity is considered, which is usually caused by rather abrupt and traumatic events; in these cases, it is not the entire health that is profoundly and lastingly affected, but certain parts of the body (e.g. skin allergies, physical injuries) or more or less persistent impairments of the psyche (e.g. depression) (3). Two other categories involve more collective damage, analysing the functioning of the economy and society (4) and the deeper consequences for the culture and identity of peoples (5).

According to the criterion of magnitude of impact, damage is also considered quantitatively and geographically, that is according to the scope of its impact. Six categories were tested, but only three of them actually showed a data connection with the damage: the damage to individuals (6) and intracommunity damage—that is within the communities directly affected, which are the local communities that are directly in or on the extractive operation area (7)—are very visible. Data also evidences extra-community damage extended beyond the immediate area, such as in areas of the province or region—without expanding onto the territory at a national scale (8). No direct damage could be observed at the national (9), transnational (10), or global (11) levels, again confirming the findings of previous work that rmr-scc is associated with global and transnational interests, but those negative consequences arising from it remain local.

Finally, the consequences were systematised according to the duration of the impact, that is the temporal perspective is in the foreground. These are: short-term damage that occurs immediately (12), medium-term damage that unfolds over a few years—up to ten years (13), and long-term damage that lasts for many years—more than ten years (14). With regard to the duration of the impact, particular attention is paid to the issue of contamination. This demonstrates a significant time lag in determining the potential damage. In the case of arsenic this is very clear:

> Since the organoleptic properties of arsenic-laced water are not particularly objectionable, i.e. it has no colour, smell or taste, individuals use the water without paying attention to possible effects. Therefore, the toxic consequences are only noticed much later. Symptoms may appear only after several years of exposure, and lesions do not become malignant until decades later. (Bundschuh et al., 2012, p. 3)

Physical illness, poisoning, and the resulting deaths are neither immediately recognisable nor apparent and can therefore be assigned to even fewer specific people. The disorders develop as if they arose naturally. They unfold very slowly.

The analysis of facts, interactions, and damage in rmr-scc makes some specific characteristics visible which can be studied as specific forms of the original phenomenon of scc.

Characteristics of Raw-Materials-Related State-Corporate Crime

Six predominant features have been observed in the deep study of these cases, in that they appeared repeatedly in related and contextual literature and research on examples of extractive industries in the global South. In the following, combining findings with a literature review, these characteristics are explained, namely actor scattering, temporal and spatial complexity of events, glocality and postcolonial interculturality in the interactions, and the nature impairment and violence evident in the social harm caused.

Actor Scattering

The scattering of actors as one of the characteristics of rmr-scc refers to the complex economic and political structures as well as to the individual, municipal, and organised foreign and local actors. In the field of white-collar crime, these actors can function as perpetrators or be regarded as victims. Many others, however, are very difficult to classify under one of the two categories. This result fits with former work on white-collar and corporate crime (Fernández Steinko, 2008, p. 38; Huisman, 2008; Huisman & van Sliedregt, 2010; Spapens, 2014, pp. 224 ff.; Sutherland, 1983) as well as with research on the criminality of the powerful (Barak, 2015; Pearce, 1976; Scheerer, 1993). The international nature of the activities was accompanied by the internationalisation of the actors, and this can be seen as a special level of diffusion. This growing internationalisation of actors is a central theme of new structures (see also Friedrichs & Rothe, 2015; Egger in Mathias, 2017; Michalowski & Kramer, 2006, p. 10), which leads also to a transformation in the relationship between society and nature (see also Hönke, 2010). As a result an increase in conflicts due to the worsening of living conditions and in relation to natural resources is given (see Altvater, 2011, p. 41 ff.; Böhm, 2016, 2019; Bran-Guzmán, 2017, p. 44; Svampa, 2017).[9] In particular, the conflicts affect the life and rights of the Indigenous people in these areas.[10]

Temporal and Spatial Complexity

Temporal complexity means the overlapping of long-term economic policy developments with medium-term interactions between institutions and short-term actions by individual actors. They all take place before, during, or after the start of a raw material operation. They intertwine synchronically and diachronically. Local complexity, on the other hand, refers to multiple scenarios in which events unfold. Business premises in industrialised countries, state offices in Latin America, large areas of mining facilities, soils, rivers and streams, community houses,

[9] Similar conflicts related to natural resources such as minerals, water, oil, and gas, as well as surfaces suitable for the production of energy (water and wind) and staple foods (cereals), have already been analysed and presented by Jarrell and Ozymy (2014), Brisman et al. (2017), Hall (2014, p. 103), and Walters (2006).

[10] For research with similar results see Boekhout and Kuijpers (2013, p. 202), Brisman et al. (2017, p. 2), Carrasco and Fernández (2009), de Carvalho et al. (2020) and Goyes et al. (2021).

and so on are the unfolding field of events, covering dimensions from on-site to trans-oceanic distances. Both characteristics in turn reflect the diffusion of the actors.

In addition to the overlapping and temporal displacement of actions, omissions, neglect, and reactions, the temporal displacement of the possibilities for dealing with the events must also be taken into account. That is, identifying which actors act before, during, or after the damaging events also defines the interests and associated opportunities that may—or may not—terminate the circumstances that support the damaging events and reduce their consequences. This is important in creating a forum for accountability for those involved in what has happened.

One could say that scc is a continuous behaviour whose timing and location are difficult to define, yet are interconnected, which is related to the idea of *macrocrime*, that political structure involved in the commission of criminal activities among violent dictatorial regimes or even—in a more updated concept—in the ongoing acceptance of human rights abuses and its active involvement in the business practices of transnational resource companies in democratic countries (Jäger, 1989; Neubacher, 2005, p. 27; Zaffaroni, 2012).[11] Social structures, state and economic interests, and, last but not least, the state itself then become a criminogenic complex scenario (see Kramer et al., 2002; Zaitch & Gutiérrez Gómez, 2015, p. 389).

Glocality

On the one hand, global economic structures (institutions, regulations, and actors) emerge from the transnational and even international level. On the other hand, the consequences of the resulting mining or oil projects are seen very precisely and visibly on territorially definable areas, that is locally, in the corresponding raw material regions, which then have immediate consequences for the local population, environment, and culture. These facts have drawn attention to the double level referred to

[11] For the terminology and the difficulties of this insufficiently defined concept, which is sometimes confused or combined with 'criminality of the powerful', see Neubacher (2005, p. 18 ss.).

here. Raw-materials-related state-corporate crime is thus simultaneously global and local in an indistinguishable way.

The global demand for raw materials and the negative local consequences are very easy to understand in the foreign trade balance. Glocality is linked to the impact distinction, as extractive industries generate large financial wealth mostly abroad on the one hand and severe environmental degradation and loss of livelihoods at the operating site on the other (see Bebbington, 2012, p. 5; Middeldorp, 2016a, 2016b p. 75). This characteristic of rmr-scc is related to the crime of maldevelopment presented and explained in detail in previous research (Böhm, 2019) as well as with the crimes of globalisation (Friedrichs, 2007; Friedrichs & Rothe, 2015; Peluso & Lund, 2011).

Postcolonial Interculturality

Interculturality, as a detected result in this research, deals with the tension between one's own culture and a foreign culture as well as between urban and rural cultures. In order to make this more visible, it is important to look at the interactions that take place beyond the economic or political arena (see Ezeonu, 2015, p. 89 f.; Kramer, 1992, 2006). The reluctance or difficulties of the source countries to regulate the markets and, at the same time, their participation in this form of (facilitated) corporate crime leads straight to the problem of postcolonial interculturality. In rmr-scc, the transnationality of the companies has a particular impact on the communication options: on the one hand, foreign actors are generally perceived as more trustworthy than local actors; on the other hand, communication between them is often problematic—especially when individuals are neglected by their own state. Isolation also increases the likelihood of victimisation (see Kirchhof, 2010).

The interaction between the population affected by new economic projects and industrial facilities on the one hand, and big business and the politicians working in capitals and ministries on the other, is often highly discriminatory and extremely disadvantageous due to lack of control. Socially speaking, large sections of the population in

Latin American countries do not have easy access to education, minorities are often marginalised or ignored, and informal jobs are common among large sections of the population. Communities with a low level of education, few employment opportunities and close contact with nature, without research opportunities or experience in bureaucratic matters, cannot really judge and decide whether the companies create new jobs or whether they respect the environment and essential food and water sources (Böhm, 2019). In this context, the indifference of the state towards the population, especially towards Indigenous peoples—and its impact especially on women—is dominant (see Svampa, 2017, p. 119 ff.).

Based on the factors mentioned, individual and social harms cannot be easily brought to justice, as judges are often part of elitist, non-Indigenous, and male groups. The social distance between them and the usual victims of extractive projects is significant.[12] Besides, the good or bad relations between political and economic actors have a direct impact on compliance or non-compliance with regulations and thus on the avoidance or non-avoidance of serious damage.[13] At the international level, these relationships can also be interpreted in terms of globalisation (Elbert, 2016, pp. 99 ff.).

Nature Impairment

Nature impairment refers to damage to nature that affects both nature as the environment (soil, water, air, flora, and fauna) and nature as a human habitat, that is, as a place where people can develop and realise their human rights.

Green criminology has already explained some key aspects related to the environment as the subject of harmful economic activity. It has provided evidence on the serious impact of this crime on the lives and rights of the most marginalised communities and on the conflicts of interest (Brisman & South, 2018; Goyes, 2015, 2019; Hall, 2014; Jarrell & Ozymy, 2014; Sheptycki, 2016), which have been even

[12] On social distance, see Edeso Natalías (2005) and Arteaga Botello and Lara Carmona (2004).

[13] In this sense, the crime of the powerful complements the concept of macrocriminality.

explained as 'environmental discrimination' (Baechler, 1999) and also as 'ecological discrimination' (Goyes, 2019). However, it is particularly relevant for rmr-scc that the environment can be viewed as a scenario, as a source, and as a 'victim'.

The privatisation of natural resources such as water and land leads to the forced displacement of communities, which in various ways affects the right to common property and thus also the communal and artisanal forms of production (see Goyes, 2015, 2019). The direct consequence is the loss of access to water and loss of food autonomy for the local population. Therefore, the profound effects on nature in terms of the environment and as an ecosystem threaten non-human and human beings in their basic physical existence (see Bran-Guzmán, 2017, p. 53; Johnson et al., 2017). When nature is affected the entire ecosystem for all living creatures, and the social and economic subsystems they create, are affected as well (Bundschuh et al., 2012).

The 'financerisation of nature' (Bruckmann, 2017) and 'commodification of nature' (Svampa, 2017, pp. 88 f.) create a general challenge in and for the global South. The 'accumulation through dispossession' (Altvater, 2011, p. 34; Harvey, 2003, pp. 137 ff.; Middeldorp, 2016a, 2016b), in a similar sense to the concept of 'necrocapitalism' (Banerjee, 2008), seems more relevant today than ever before.

Industries that pollute the environment with their negative effects are not only relocated to countries in the global South because of the raw materials available locally. They are often settled there precisely because the lack of rules, lack of control, and indifference towards the population living there make it easier to deposit toxic waste and residual products in these 'pollution havens' (Cheng & Shi, 2017, pp. 295 ff.; see Doytch & Uctum, 2016, p. 584; Eskeland & Harrison, 2003, p. 1). The extractive industry paradigmatically combines the extraction of raw materials with the contamination of their sources, the environment, and the people living there.

Violence

Violence is a feature of rmr-scc in its invisible and visible expressions, not only because of it but also as an economic tool.

The specific violence of extractive activities as damage of high severity can be explained in different ways. Nancy Peluso and Christian Lund (2011) describe four 'land control mechanisms' they observed: *enclosure* by establishing physical or institutional fences around certain resources, *territorialisation* by the creation of institutional alliances and commitments, *legalisation* when through an administrative decision common property converts landholders into trespassers, and *force and violence* (Middeldorp, 2016a, 2016b p. 73 ff; Peluso & Lund, 2011, pp. 667 ff.). Although only the last one is called violence, the four mechanisms refer to invisible and visible violence according to Johan Galtung (1969, 1996, 2003; see also Böhm, 2019).

Eventually, violence in the usual sense of the word—direct, physical, and individual—is the most visible form of the mechanisms of territorial claim and control. This is about the violence, including militarisation and/or the use of terror, which is observed, for example, in the action against the local population and in the reaction of state-owned companies to social movements when they raise their voices collectively, demand their rights, resist serious uncontrolled extractive projects, or demand reforms to change conditions (see Goyes, 2015; Middeldorp, 2016a, 2016b pp. 73 ff.; Peluso & Lund, 2011, pp. 675 ff.).[14]

In this context, physical (direct) violence perpetrated by different actors can be systematised in the analysis and classified under four different types: (1) physical violence by individuals and communities affected by and opposed to economic projects; (2) physical violence by corporations, businessmen, or entrepreneurs to improve the economic profits made by expanding their territory; (3) physical violence by the state to defend the established order and protect specific economic activity. In addition, common forms of state violence are the criminalisation and suppression of social protests (4) and physical violence by

---

[14] See *Honduras: Mártires de la tierra*, documentary film, Dir. Julián del Olmo, Spain: Pueblo de Dios TVE, 2013.

those working in the newly added facilities (see Böhm, 2019, pp.1 57 ff., 165 ff.). These four categories of physical violence are not only a consequence of deregulated business operations and their conditions, but they go far beyond the existing close connections and can be termed 'chains of violence' (Svampa, 2017, p. 119).[15]

Visible (direct) and invisible (indirect) violence, therefore, tend to increase after the opening of new industrial or mining plants in a given area. This is very evident in a specific population group: women. As an impact of rmr-scc, violence against women—in its most diverse forms—is a salient feature. In addition to the physical violence they are subjected to and suffer from, women are often deprived of their traditional roles while at the same time are excluded from the opportunity to work for the company (see JASS, 2012; WRM, 2003). There is violence against women, but it is usually not prosecuted (Carlsen, 2012; RLMDDSA, 2017).

Violence and violent activities are not usually associated in the collective imagination with the economic activity of a transnational corporation. However, they are more common than thought and very damaging physically, both individually and collectively,[16] and this is directly related to this feature of rmr-scc.

Conclusion

The nature of extractive activity in the twenty-first century is characterised by a long tradition of geopolitical imbalances between regions. Some world regions own and export their natural resources and others import them as commodities for further processing. It has also been noted that the current forms of these activities have specific characteristics and that one of these characteristics is deregulation—a highly criminogenic global phenomenon. International resource extraction projects act as catalysts of deregulation, accelerating the deepening of legal and

[15] Even the Inter-American Commission on Human Rights in Costa Rica has pointed out the violent situations at various levels (IACHR, Doc 47/15, Para. 297 ss.; IACHR: Doc 49/15).

[16] See Hills (1987) ('corporate violence'); Ruggiero (1997, 2007a, 2007b) ('dirty economies'); Albrecht (2007) ('economy of violence', 'Gewaltökonomie').

institutional orders and political decisions that, despite their potential negative consequences, prioritise over-exploitation at the expense of alternative less harmful project implementations. On the one hand, the extractive activity is developed from projects that need higher investment, higher risks, and longer deadlines compared to other industries. On the other hand, the time span in which the extractive activities are carried out is relatively short when viewed against the background of traditional cultures and societies. This contrast highlights the short-term expectations of the economic actors who carry out such ventures and the political actors who promote or facilitate such projects. The natural wealth that has always existed, and the ancestral societies and cultures within it, are undoubtedly more enduring than the instantaneity of mining activity or oil extraction, which are exhausted in a few years. However, the damage in turn has a more permanent impact on the natural environment, given the relatively short period of extractive activity.

Many different interests and interactions coincide with the onset of extractive activities, namely the increased needs of the market, the all-too-ready provision of the riches by indifferent politicians and lax state policies, the qualitatively and quantitatively aggressive methods of extraction, and the ruthless discrimination against local people affected by the activities. All this makes these economic complexes a phenomenon that inevitably draws attention to two important groups of actors: economic actors and political actors. These actors accept possible harm as side effects. The state is fully involved in projects of the raw materials industry and their consequences.

Adopting the concept of scc and related theoretical approaches contributes to understanding that corporations would not be able to advance their projects without the active participation and interest or support of the state. Without the involvement of state actors, corporations could neither get the privileges they have nor inflict the harm they do. All of this is only possible because the state is directly involved in business interests and can also pursue its own interests through large-scale extractive activities, which is why the state does not fulfil its control and sanction obligations.

References

Albrecht, H.-J. (2007). Internationale Kriminalität, Gewaltökonomie und Menschenrechtsverbrechen: Antworten des Strafrechts. *Internationale Politik und Gesellschaft, 2*, 153–169.

Alonso, S. A. (2020). Caso 'Lote 8'. Empresa canadiense en Guatemala. In M. L. Böhm (Dir.), *Empresas transnacionales, recursos naturales y conflicto en América Latina. Para una visibilización de la violencia invisible* (pp. 177–185). Universidad de Buenos Aires.

Altvater, E. (2011). *Los límites del capitalismo. Acumulación, crecimiento y huella ecológica.* Mardulce.

Arteaga Botello, N., & Lara Carmona, V. L. (2004). Violencia y distancia social: una revisión. *Papeles de Población CIEAP/UAEM, 40*(April/June), 169–191.

Aulette, J. R., & Michalowski, R. (1993). Fire in Hamlet: A case study of state-corporate crime. In K. D. Tunnell (Ed.), *Political crime in contemporary America: A critical approach* (pp. 171–206). Garland.

Aulette, J. R., & Michalowski, R. (2006). The fire in Hamlet. In R. J. Michalowski & R. C. Kramer (Eds.), *State-corporate crime: Wrongdoing at the intersection of business and government* (pp. 45–66). Rutgers University Press.

Baechler, G. (1999). Violence through environmental discrimination: Causes, Rwanda arena, and conflict model. In *Social indicators research series* (Vol. 2). Springer-Science + Business Media, B.V.

Banerjee, S. B. (2008). Necrocapitalism. *Organization Studies, 29*(12), 1541–1563.

Barak, G. (Ed.). (2015). *The Routledge international handbook of the crimes of the powerful.* Routledge.

Barak, G. (2016). Alternatives to high-risk securities fraud control: Proposing structural transformation in an age of financial expansionism and unsustainable global capital. *Crime Law and Social Change, 66*, 131–145.

Barak, G. (2017). *Unchecked corporate power. Why the crimes of multinational corporations are routinized away and what we can do about it*. Routledge.

Bebbington, A. (2012). Extractive industries, socio-environmental conflicts, and political economic transformations in Andean America. In A. Bebbington (Ed.), *Extractive industries, social conflict and economic development: Evidence from South America* (pp. 3–26). Routledge.

Boekhout van Solinge, Tim/Kuijpers, K. (2013). The amazon rainforest. A green criminological perspective. In South, N./Brisman, A. (Eds.), *Routledge International Handbook of Green Criminology* (pp. 199–213). Routledge.

Böhm, M. L. (2016). Transnational corporations, human rights violations and structural violence in Latin America: A criminological approach. *Kriminologisches Journal, 4*, 272–293.

Böhm, M. L. (2019). *The crime of maldevelopment: International deregulation and violence in the global south*. Routledge.

Bran-Guzmán, E. (2017). Conflictividad socioambiental en Centroamérica. Una década de rearticulación y movilización social y política. *Argumentos, 30*(83), 43–68.

Brasil, L. M., & Pires, V. (2017). O que a lama nos deixou: reflexões sobre a tragédia de Mariana, a mídia ea mineração no Brasil. *Chroniques des Amériques, 17*(3a), 1–10.

Brisman, A., & South, N. (2018). Environment, conflict and profit: Harmful resource exploitation and questionable revenue generation. In T. Spapens, R. White, D. van Uhm, & W. Huisman (Eds.), *Green crimes and dirty money* (pp. 19–41). Routledge.

Brisman, A., South, N., & White, R. (2017). Toward a criminology of environment—Conflict relationships. In A. Brisman, N. South, & R. White (Eds.), *Environmental crime and social conflict: Contemporary and emerging issues* (pp. 1–38). Ashgate.

Bruckmann, M. (2017, February 13). La financiarización de la naturaleza y sus consecuencias geopolíticas. *Revista Diálogos del Sur*. https://dialogosdelsur.operamundi.uol.com.br/medio-ambiente/57341/la-financiarizacion-de-la-naturaleza-y-sus-consecuencias-geopoliticas. Accessed 2 Nov 2021.

Bundschuh, J., Litter, M. I., Parvez, F., Román-Ross, G., Nicolli, H. B., Jean, J.-S., Liu, C.-W., López, D., Armienta, M. A., Guilherme, L. R. G., Gómez Cuevas, A., Cornejo, L., Cumbal, L., & Toujaguez, R. (2012). One century of arsenic exposure in Latin America: A review of history and occurrence from 14 countries. *Science of the Total Environment, 429*, 2–35.

Bussmann, K. D. (2010). Wirtschaftskriminologie in Aufbruch. *Monatschrift für Kriminologie und Strafrechtsreform, 93*(5), 343–345.

Cálix, M. (2017, October 4). El Valle de la desesperanza. *Contracorriente*. https://contracorriente.red/2017/10/05/el-valle-de-la-esperanza/. Accessed 21 June 2021.

Carlsen, L. (2012, December 10). Matar a la mensajera: Crece la violencia contra las defensoras de derechos humanos. *Pensamiento Crítico*. www.pensamientocritico.org/laucar0113.htm. Accessed 2 Nov 2021.

Carrasco, A., & Fernández, E. (2009). Estrategias de resistencia indígena frente al desarrollo minero. La comunidad de Likantatay ante un posible traslado forzoso. *Estudios Atacameños, 38*, 75–92.
Cheng, D., & Shi, M. (2017). Multiple environmental policies and pollution haven hypothesis: Evidence from China's polluting industries. *Journal of Cleaner Production, 141*, 295–304.
Crystal, V., Imai, S., & Maheandiran, B. (2014). Access to justice and corporate accountability: A legal case study of HudBay in Guatemala. *Osgoode Legal Studies Research Paper Series, 73*, 1–26.
Cunneen, C. (2015). The race to defraud—State crime and the immiseration of Indigenous people. In W. J. Chambliss & C. J. Moloney (Eds.), *State crime: Critical concepts in criminology. Vol. II: Varieties of state crimes* (pp. 658–671). Routledge.
Cunneen, C., & Tauri, J. (2017). Survival, dignity and wellbeing: Indigenous human rights and transformative approaches to justice. In L. Weber (Ed.), *The Routledge international handbook of criminology and human rights* (pp. 429–439). Routledge.
da Silva, G. A., Teixeira Boava, D. L., & Felício Macedo, F. M. (2017). Refugiados de Bento Rodrigues: o desastre de Mariana. *MG. Revista Pensamento Contemporâneo em Administração, 11*(2), 63–81.
Dammert, B. J. L. (2019). The growth of oil palm in the Peruvian Amazon: Deforestation and land trafficking. In A. Chirif (Ed.), *Peru: Deforestation in times of climate change* (pp. 53–69). International Work Group for Indigenous Affairs.
de Carvalho, S., Goyes, D. R., & Vegh Weis, V. (2020). Politics and indigenous victimization: The case of Brazil. *The British Journal of Criminology, 61*(1), 251–271. https://doi.org/10.1093/bjc/azaa060%JTheBritishJournalofCriminology
Donziger, S., Garr, L., & Page, A. M. (2010). The clash of human rights and BIT investor claims: Chevron's abusive litigation in Ecuador's Amazon. *Human Rights Brief, 17*(2), 8–15.
Doytch, N., & Uctum, M. (2016). Globalization and the environmental impact of sectoral FDI. *Economic Systems, 40*, 582–594.
Edeso Natalías, V. (2005). La distancia social y su importancia en la interacción: propuesta para su estudio en clase de ELE. *ASELE, Actas, XVI*, 247–257.
Eisenberg, U., & Kölbel, R. (2017). *Kriminologie. 7. Auflage.* Mohr Siebeck.
Elbert, C. A. (2016). *Criminología Latinoamericana. Identidad, realidad social y Estado.* Res Pública.

Eskeland, G. S., & Harrison, A. E. (2003). Moving to greener pastures? Multinationals and the pollution haven hypothesis. *Journal of Development Economics, 70*, 1–23.

Ezeonu, I. C. (2015). Capital and catharsis in the Nigerian petroleum extraction industry: Lessons on the crimes of globalization. In G. Barak (Ed.), *The Routledge international handbook of the crimes of the powerful* (pp. 89–104). Routledge.

Ezeonu, I. C. (2018). *Market criminology: State-corporate crime in the petroleum extraction industry*. Routledge.

Fernández Steinko, A. (2008). *Las pistas falsas del crimen organizado: Finanzas paralelas y orden internacional*. Catarata.

Fernández Steinko, A. (2013). Una visión de conjunto de la delincuencia contemporánea. In A. Fernández Steinko (Ed.), *Delincuencia, Finanzas y Globalización* (pp. 15–129). Centro de Investigaciones Sociológicas.

Friedrichs, D. O. (2006). *Trusted criminals: White collar crime in contemporary society*. Thomson Wadsworth.

Friedrichs, D. O. (2007). Transnational crime and global criminology: Definitional, typological, and contextual conundrums. *Social Justice, 34*(2), 4–18.

Friedrichs, D. O., & Rothe, D. L. (2015). *Crimes of globalization*. Routledge.

Galeano, E. (1984). *Las venas abiertas de América Latina*. Siglo XXI.

Galindo da Fonseca, P., & Galindo da Fonseca, I. (2016). Brazil's greatest environmental catastrophe—Samarco Fundão Tailings Dam. *Environmental Policy and Law, 46*(5), 334–337.

Galtung, J. (1969). Violence, peace, and peace research. *Journal of Peace Research, 6*(3), 167–191.

Galtung, J. (1996). *Peace by peaceful means: Peace and conflict, development and civilization*. Sage.

Galtung, J. (2003). *Violencia Cultural—Documento de trabajo 14*. Centro de Investigación por la Paz.

Godson, R. (Ed.). (2003). *Menace to society: Political-criminal collaboration around the world*. Transaction Publishers.

Gómez Lende, S. (2017). Minería del litio y acumulación por desposesión: El caso de Salar del Hombre Muerto (1999–2016). *Estudios Geográficos, Río Claro, 15*(1), 157–183.

Gorenstein, S. (2016). *Empresas transnacionales en la agricultura y la producción de alimentos en América Latina y el Caribe*. Nueva Sociedad/Friedrich Ebert Stiftung.

Goyes, D. (2015). Land uses and conflict in Colombia. In A. Brisman, N. South, & R. White (Eds.), *Environmental crime and social conflict: Contemporary and emerging issues* (pp. 265–283). Ashgate.

Goyes, D. (2019). *Southern green criminology: A science to end ecological discrimination.* Emerald Publishing Limited.

Goyes, D., & South, N. (2016a). Land-grabs, biopiracy and the inversion of justice in Colombia. *The British Journal of Criminology, 56*(3), 558–577.

Goyes, D., & South, N. (2016b). The injustices of policing, law and multinational monopolisation in the privatisation of natural diversity: Cases from Colombia and Latin America. In D. R. Goyes, H. Mol, A. Brisman, & N. South (Eds.), *Environmental crime in Latin America: The theft of nature and the poisoning of the land*. Palgrave.

Goyes, D. R., & South, N. (2019). Between 'conservation' and 'development': The construction of 'protected nature' and the environmental disenfranchisement of indigenous communities. *International Journal for Crime, Justice and Social Democracy, 8*(3), 89–104.

Goyes, D. R., Abaibira, M. A., Baicué, P., Cuchimba, A., Ñeñetofe, D. T. R., Sollund, R., South, N., & Wyatt, T. (2021). Southern green cultural criminology and environmental crime prevention: Representations of nature within four Colombian Indigenous communities. *Critical Criminology, 29*(3), 469–485. https://doi.org/10.1007/s10612-021-09582-0

Goyes, D. R., Komatsubara, O., Droz, L., & Wyatt, T. (2022). Green criminological dialogues: Voices from Asia. *International Journal for Crime, Justice and Social Democracy, 11*(1), 1–10.

Goyes, D. R., Mol, H., Brisman, A., & South, N. (Eds.) (2017). *Environmental crime in Latin America: The theft of nature and the poisoning of land*. Palgrave Macmillan.

Goyes, D. R., Sollund, R., & South, N. (2019). Guest editors' introduction towards global green criminological dialogues: Voices from the Americas and Europe. *International Journal for Crime, Justice and Social Democracy, 8*(3), 1–5.

Gudynas, E. (2015). *Derechos de la Naturaleza, Ética biocéntrica y políticas ambientales.* Tinta Limón.

Hall, M. (2014). Victims of environmental crime: Routes for recognition, restitution and redress. In T. Spapens, R. White, & M. Kluin (Eds.), *Environmental crime and its victims: Perspectives within green criminology* (pp. 103–118). Ashgate.

Harvey, D. (2003). *The new imperialism.* Oxford University Press.

Henry, S. (1991). The informal economy: A crime of omission by the state. In G. Barak (Ed.), *Crimes by the capitalist state: An introduction to state criminality* (pp. 253–270). State University of New York Press.

Hillebrand, E. (2016). *Branchenanalyse Rohstoffindustrie, Reihe Study Nr. 315*. Hans Böckler Stiftung.

Hills, S. L. (Ed.). (1987). *Corporate violence: Injury and death for profit*. Rowman & Littlefield.

Hönke, J. (2010). New political topographies: Mining companies and indirect discharge in Southern Katanga (DRC). *Politique Africaine, 4*(120), 105–127.

Hoogenboom, B. (2010). *The governance of policing and security: Ironies, myths and paradoxes*. Palgrave Macmillan.

Huisman, W. (2008). Corporations and international crimes. In A. Smeulers & R. Haveman (Eds.), *Supranational criminology: Towards a criminology of international crimes* (pp. 181–211). Intersentia.

Huisman, W., & van Sliedregt, E. (2010). Rogue traders: Dutch businessmen, international crimes and corporate complicity. *Journal of International Criminal Justice, 8*(3), 803–828.

IACHR (Inter-American Commission of Human Rights). (2015a). *Informe de la Comisión Interamericana de Derechos Humanos sobre Criminalización de defensoras y defensores de derechos humanos*, OEA/Ser.L./V/II. Doc. 49/15, 31/12/2015a.

IACHR (Inter-American Commission of Human Rights). (2015b). *Informe de la Comisión Interamericana de Derechos Humanos sobre Pueblos indígenas, comunidades afrodescendientes y recursos naturales: protección de derechos humanos en el contexto de actividades de extracción, explotación y desarrollo*, OEA/Ser.I./V/II. Doc. 47/15, 31/12/2015b.

IDAMHO (Instituto de Derecho Ambiental de Honduras). (2012). *La mina San Martín en el Valle de Siria. Exploración, explotación y cierre: impactos y consecuencias*. IDAHO/OXFAM.

Jäger, H. (1989). *Makrokriminalität: Studien zur Kriminologie kollektiver Gewalt*. Suhrkamp.

Jarrell, M., & Ozymy, J. (2014). Communities as victims of environmental crime: Lessons from the field. In T. Spapens, R. White & M. Kluin (Eds.), *Environmental crime and its victims: Perspectives within green criminology* (pp. 249–261). Ashgate.

JASS. (2012). Maya K'Iche' leader says no to violence against women. *JustAssociates*. https://justassociates.org/en/womens-stories/maya-kiche-leader-says-no-violence-against-women. Accessed 15 June 2017.

Johnson, H., South, N., & Walters, R. (2017). Environmental crime, human rights and green criminology. In L. Weber (Ed.), *The Routledge international handbook of criminology and human rights* (pp. 210–219). Routledge.
Kimerling, J. (1991). *Amazon crude*. Natural Resources Defense Council.
Kimerling, J. (1994). The environmental audit of Texaco's Amazon oil fields: Environmental justice or business as usual? *Harvard Human Rights Journal, 7*, 199–224.
Kimerling, J. (2006). Indigenous peoples and the oil frontier in Amazonia: The case of Ecuador, Chevron Texaco, and Aguinda v. Texaco. *New York University Journal of International Law and Politics, 38*, 413–664.
Kirchhof, G. F. (2010). History and a theoretical structure of victimology. In S. G. Shoham, P. Knepper, & M. Kett (Eds.), *International handbook of victimology* (pp. 96–123). Taylor & Francis.
Kölbel, R. (2019). „Revolving Doors" als institutionelle Korruption im legislatorischen Feld. In J. Krüper (Hrsg.), *Die Organisation des Verfassungsstaats. Festschrift für Martin Morlok zum 70. Geburtstag* (pp. 261–276). Mohr Siebeck.
Kramer, R. C. (1992). The space shuttle challenger explosion. In K. Schlegel & D. Weisburd (Eds.), *White collar crime reconsidered*. Northeastern University Press.
Kramer, R. C. (2006). The space shuttle *challenger* explosion. In R. J. Michalowski & R. C. Kramer (Eds.), *State-corporate crime: Wrongdoing at the intersection of business and government* (pp. 27–44). Rutgers University Press.
Kramer, R. C., & Michalowski, R. J. (2006). The original formulation. In R. J. Michalowski & R. C. Kramer (Eds.), *State-corporate crime: Wrongdoing at the intersection of business and government* (pp. 18–26). Rutgers University Press.
Kramer, R. C., Michalowski, R. J., & Kauzlarich, D. (2002). The origins and development of the concept and theory of state-corporate crime. *Crime & Delinquency, 48*(2), 263–282.
Mathias, M. (2017, September 19). Matopiba: na fronteira entre a vida e o capital. *EPSJV/Fiocruz*. http://www.epsjv.fiocruz.br/noticias/reportagem/matopiba-na-fronteira-entre-a-vida-e-o-capital. Accessed 3 Nov 2021.
Matthews, R. A., & Kauzlarich, D. (2000). The crash of ValuJet Flight 592: A case study in state-corporate crime. *Sociological Focus, 33*(3), 281–298.

Méndez Gutiérrez, L. (2013, October). Guatemala: Abriendo brecha en la búsqueda de la justicia – las valientes mujeres q'eqchís. *visionews.net*. http://www.visionews.net/es/guatemala-abriendo-brecha-en-la-busqueda-de-justicia-las-mujeres-valientes-q%C2%B4eqchis/. Accessed 16 June 2021.

Michalowski, R. J., & Kramer, R. C. (2006). The critique of power. In R. J. Michalowski & R. C. Kramer (Eds.), *State-corporate crime: Wrongdoing at the intersection of business and government* (pp. 1–17). Rutgers University Press.

Middeldorp, N. (2016a). "La vida no tiene precio": La oposición a la minería en Honduras. *Anuario de Estudios Centroamericanos, 42*, 63–92.

Middeldorp, N. (2016b). Minería, resistencia y represión en Honduras: entre la ley y la impunidad. *Cuadernos de Antropología, 26*(2), 69–89.

Neubacher, F. (2005). *Kriminologische Grundlagen einer internationalen Strafgerichtsbarkeit: politische Ideen- und Dogmengeschichte, kriminalwissenschaftliche Legitimation, strafrechtliche Perspektiven.* Mohr Siebeck.

Opel, A. (2016). *Ausländische Agrarinvestitionn: "Land-Grabbing" im Spannungsfeld zwischen Menschenrechtsschutz und Investitionsschutzrecht, Schriften zum Völkerrecht, Bd. 221.* Duncker & Humblot.

Ortiz, G. (2011, March 31). Los árboles tiemblan en la selva de Texaco. *Rebelion* (Orig. in *Tierramérica*, 2011.03.26). https://rebelion.org/los-arboles-tiemblan-en-la-selva-de-texaco/. Accessed 3 Nov 2021.

Palacios, J.-L., Calvo, G., Valero, A., & Valero, A. (2018). Exergoecology assessment of mineral exports from Latin America: Beyond a tonnage perspective. *Sustainability, 10*(723), 1–18.

Parra, R. (Ed.). (2014). *El Camino por la Justicia – Victimización y Resistencia de mujeres indígenas y campesinas en Guatemala y Colombia.* Universidad Javeriana/Universidad de Antioquia.

Pearce, F. (1976). *Crimes of the powerful: Marxism, crime and deviance*. Pluto Press.

Peluso, N. L., & Lund, C. (2011). New frontiers of land control: Introduction. *The Journal of Peasant Studies, 38*(4), 667–681.

Pieth, M. (2019). *Goldwäsche: Die schmutzigen Geheimnisse des Goldhandels.* Salis.

Rachman, G. (2016). *Easternisation: War and peace in the Asian century.* Penguin.

Raskovsky, R. (2020). Técnicas de neutralización y Arbitraje Internacional de Inversiones. In K. Ambos & M. L. Böhm (Eds.), *Empresas transnacionales y graves violaciones de Derechos Humanos en América Latina* (pp. 143–190). Tirant lo blanch/CEDPAL.

RLMDDSA (Red Latinoamericana de Mujeres Defensoras de los Derechos Sociales y Ambientales). *Letter Asking for the Stopping of Violent Extractive Activities in Latin America, 2017*. https://www.salvalaselva.org/peticion/1044/mujeres-denuncian-que-proyectos-extractivos-generan-violencia. Accessed 10 Nov 2021.

Ruggiero, V. (1996). War markets: Corporate and organized criminals in Europe. *Social and Legal Studies, 5*, 5–20.

Ruggiero, V. (1997). Criminals and service providers: Cross-national dirty economies. *Crime, Law & Social Change, 28*, 27–38.

Ruggiero, V. (2007a). Guerra, delito, imperio y cosmopolitanismo. In M. L. Böhm & M. H. Gutiérrez (Eds.), *Políticas de Seguridad: Peligros y desafíos para la criminología del nuevo siglo* (pp. 317–334). Editores del Puerto.

Ruggiero, V. (2007b). It's the economy, stupid! Classifying power crime. *International Journal of the Sociology of Law, 24*(4), 163–177.

Russau, C. (2016, November 19). Der *Dammbruch bei Mariana – eine politische, soziale und ökologische Katastrophe mit Ansage*. Presentation at Roundtable Brazil KoBra. https://www.kooperation-brasilien.org/de/veranstaltungen/runder-tisch-brasilien/rtb-2016-1/1911_forum2/bergbau-der-dammbruch-bei-mariana-2013-eine-politische-soziale-und-oekologische-katastrophe-mit-ansage. Accessed 21 Apr 2020.

Sandoval Vinelli, C. D. (2018). Criminología Verde y el Agua: Estudio aplicado del caso de Zaruma Ecuador. *M+A. Revista electrónic@ de Medio Ambiente, 19*(2), 238–256.

Scheerer, S., et al. (1993). Kriminalität der Mächtigen. In G. Kaiser (Ed.), *Kleines kriminologisches Wörterbuch* – (3rd ed., pp. 246–249). C.F. Müller.

Sheptycki, J. (2016). Is green criminology paradigm-breaking? Some reflections on hydrocarbon and resource extraction, crime and criminological thinking. In A. C. Spapens (Ed.), *Environmental crime in transnational context* (pp. 67–82). Routledge.

Spapens, T. (2014). Invisible victims: The problem of policing environmental crime. In T. Spapens, R. White, & M. Kluin (Eds.), *Environmental crime and its victims: Perspectives within green criminology* (pp. 221–236). Ashgate.

Stańczak, D. (2017). State-corporate crime and the case of BT cotton: On the production of social harm and dialectical process. *State Crime Journal, 6*(2), 214–240.

Sutherland, E. (1983). *White collar crime: The uncut version*. Yale University Press.

Svampa, M. (2017). *Del cambio de época al fin de ciclo. Gobiernos progresistas, extractivismo y movimientos sociales en América Latina*. Edhasa.

Svampa, M., & Viale, E. (2014). *Maldesarrollo*. Katz.
Torres Funes, A. (2016, September 27). Valle de Siria: Oro, pobreza y resistencia. *El Pulso HN*. http://elpulso.hn/valle-de-siria-oro-pobreza-y-resistencia/. Accessed 10 Nov 2021.
Torres Wong, M. (2018). *Natural resources, extraction and indigenous rights in Latin America: Exploring the boundaries of environmental and state corporate crime in Bolivia, Peru and Mexico*. Routledge.
Trucchi, G. (2014, September 7). Actividad minera en región de Honduras deja rastro de enfermedades, destrucción ambiental y desempleo. *Nicaragua y más*. https://nicaraguaymasespanol.blogspot.com.ar/2014/09/actividad-minera-en-region-de-honduras.html. Accessed 10 Nov 2021.
Walters, R. (2006). Crime, bio-agriculture and the exploitation of hunger. *British Journal of Criminology, 46*, 26–45.
White, R., & South, N. (2013, November). *The future of green criminology: Horizon scanning and climate change*. Paper presented at the American Society of Criminology Conference, Atlanta.
WRM. (2003, June 30). Los impactos de la minería sobre las mujeres. *Boletín WRM*. https://wrm.org.uy/es/?s=los+impactos+de+la+miner%C3%ADa+en+las+mujeres. Accessed 10 Nov 2021.
Zabalo, P. (2008). Los acuerdos internacionales sobre inversión, otro obstáculo para el desarrollo de América Latina. *Gestión en el Tercer Milenio, 11*(22), 27–39.
Zaffaroni, E. R. (2012). *Crímenes de masa* (2nd ed.). Ediciones Madres de Plaza de Mayo.
Zaitch, D., & Gutiérrez Gómez, L. (2015). Mining as state-corporate crime: The case of AngloGold Ashanti in Colombia. In G. Barak (Ed.), *The Routledge international handbook of the crimes of the powerful* (pp. 386–397). Routledge.

3

Mass Extraction and Green Crime Victimization in Turkey

Halil Ibrahim Bahar

Introduction

Our natural environment is suffering from the systematic threat of deliberate destruction. With climate change, the ruinous damaging of ecosystems, the complete removal of species, and the pollution of air, land, and water, the sustainability of the planet is under great threat. This widespread, severe, and systematic mass destruction of the environment and its ecosystems is also called ecocide. There is a direct relationship between the proliferation of corporations and ecocide (White, 2020). International campaigners propose a global recognition of the pressing nature of this ecocide (Fletcher, 2021; Higgins et al., 2013). A major report released by the United Nations Food and Agriculture Organization (FAO, 2021, p. 1) warned that "the interconnected systems of land, soil and water are stretched to the limit". The report stresses that

H. I. Bahar (✉)
Freelance, Ankara, Turkey
e-mail: hibahar@hotmail.com

D. R. Goyes (ed.), *Green Crime in the Global South*, Palgrave Studies in Green Criminology, https://doi.org/10.1007/978-3-031-27754-2_3

"farming systems are becoming polarized. Large commercial holdings now dominate agricultural land use, while fragmentation of smallholders concentrates subsistence farming on lands susceptible to degradation and water scarcity" (p. 1). As a solution, a more inclusive and adaptive method of land and water governance is suggested in the report.

The 2018 Environmental Performance Index, published by Yale University, ranks Turkey 108th out of 180 countries (Wendling et al., 2018). It can be theorized that there is a link between the quality of democracy and green victimization in Turkey. In the international reports, serious concerns have been raised about the deterioration of human rights, the rule of law, freedom of expression, the fight against corruption, and many other elements of a healthy social and political life in the country (Human Rights Watch, 2021; United States Embassy and Consulates in Turkey, 2020). Separation of powers, a *sine qua non* of democracy, is nonexistent. The check and balance system is nonexistent; the judiciary is partial. The police are not accountable to the public and the law. Patterns of inequality have been increasing. The independent media has been silenced. All political opposition, dissidents, and resistance movements, including resistance to green crime, are marginalized and met with the brutality of security forces and other forms of ill-treatment. Social institutions are not inclusive, and since it came to the power in 2002, the ruling Justice and Development Party (AKP) applies the divide and rule tactic as part of establishing and maintaining national hegemony.

Interest groups, especially corporate companies, play a role in the definition of what is a green crime (Lynch & Stretesky, 2003). Denying the harms of green crime in Turkey, the AKP says "nothing is lost or stolen from the coffers of the state and nobody is harmed". The party proposes that extractive business brings jobs and revenue to the economy. In reality, these projects cause irreversible damage to the ecosystem. People living in rural areas are left with no option other than abandoning lives rooted in agriculture and migrating to urban areas where they face many social, economic, and political challenges. For instance, in May 2014, an explosion and fire broke out at a coalmine in the town of Soma, killing 301 men. The Soma Mining Disaster (SMD) shows signs that Turkey's natural resources are open to harmful exploitation by national and international corporations—all facilitated by an undemocratic atmosphere. Green crime and victimization in Turkey have increased conflicts over environmental justice at local and national levels. One of the important reasons behind the development of environmental justice conflict is, as Aydin (2019, p. 1) points out, "not only due to unequal distribution of economic and ecological costs and benefits but also due to lack of participation in decision-making and recognition of rights and identities".

The main aim of this chapter is to investigate the impact of the extractive industry on green crime victimization in Turkey. I will use as case studies the 2014 SMD and the 2021–2022 Eastern Black Sea (EBS) region of the Ikizdere Stone Pit Project (ISPP). I will also analyze the resistance movements working on those cases. Some courageous local and national activists along with victims of green crime, mostly women, take part in collective resistance movements in efforts to try to protect their livelihood and right to live. In this study I will ask who profits from the mass extraction of Turkey's natural resources and who challenges green crime and victimization. Taking a Southern approach to green crime, this study shows how interactions between international and national corporations, politicians, and powerful elites—under an undemocratic atmosphere—facilitate dynamics that destroy the environment and the livelihood of rural people. These dynamics also benefit the rich and powerful while depriving the rest of their fundamental economic, political, and social rights to well-being.

The main argument of this chapter is that corruption and lack of democracy increases the risk of green victimization. While government-backed, national, and international corporations earn a significant amount of wealth from extractive and energy projects, green victimization is ignored. The chapter underscores how the political and economic climate in Turkey puts people at an increased risk of becoming victims of green crime and finding themselves in confrontation with security forces. The content of the chapter is as follows. First, a brief review will be presented on the Southern green criminology debates. Then, the political, economic, and ecological context of Turkey will be discussed. Later, two cases, the SMD and the ISPP, will be put forward. The study concludes by suggesting that unless substantial, structural, political, and economic changes are made, the vulnerability of people to green crime and the conflict between people and the security forces will continue. I base my assertions on the available literature, official reports, and other online material.

Debates Within Southern Green Criminology

Environmental destruction is not always considered a crime. For example, many governments allow deforestation or the dumping of toxic waste. Yet, green crime happens more often and is more widespread than street crime (Lynch et al., as cited in Lynch, 2020, p. 51). Scholars pay attention to what they call 'green harms' that produce injury and victimization, even though the law might allow them (Jarrell et al., 2014, p. 423). Although a green perspective into criminology can be dated back to the 1990s, some earlier literature on the sociology of deviance and criminology, social problems, and political economy had already dealt with environmental issues and the abuse of nature (Goyes & South, 2017; Lynch, 1990; South, 1998).

Green criminology is interested in environmental harm and deals with the study of crime, harm, and injustice related to the environment and species other than humans (South, 2014, p. 8). Southern criminology debates from 'the periphery' have begun to widen (Brisman & South, 2019; Carrington, 2021; Carrington et al., 2016, 2018a, 2018b; Connell, 2007; Goyes, 2019a; Goyes & South, 2017; Jarrell et al., 2014; South, 1998; White & Eman, 2020), and many others have been major catalysts in promoting Southern criminology and Southern green criminology. One of the main reasons behind the rise in Southern criminology is that the production of 'knowledge' in the global North has been critiqued as Anglophone and Eurocentric but with a presumed universal application (Connell, 2007). Therefore, the aim of Southern criminology is to redress the imbalance in criminological knowledge by clearing it of the biases present in hegemonic Northern views (Carrington et al., 2016, 2018a, 2018b).

The metaphor 'North–South' or 'center–periphery' is conceptualized for the unequal, economic, political, and knowledge power relations in which the North enjoys economic, political, and intellectual domination over the South (Goyes, 2019a, 2021). The North is rich in financial capital. This does not mean that it also should be richer than the South in producing knowledge, even though the North has disproportionate means of producing intellectual knowledge, such as universities, journals, grants, and other instruments. The advocates of Southern criminology argue that the South must have its own theoretical tools as well—without ignoring the tools that the North can provide (Carrington et al., 2016, 2018a, 2018b).

Scholars from the Southern criminology school (Carrington et al., 2016, 2018a, 2018b; Carrington, 2021; Goyes, 2019a; Goyes & South, 2017) propose that global North–South inequalities in knowledge production are not just about basic infrastructure, certification, exploitation, and colonialism, but that they also have injurious consequences. The green crime debate has, for decades, been made peripheral in Northern criminology. Southern criminology seeks to redress this gap "as a salve for the biases of metropolitan thinking" (Hughes, as cited in Carrington, 2021).

One of the distinguished pioneers of Southern criminology, Carrington (2021), defines the branch as an intellectual, political, and theoretical project. Southern criminology argues that spatial, geopolitical, and social differences are lost within the imperial narrative of the time. Southern criminology criticizes the way of thinking that defines the North as 'developed' and 'first world'. This may lead to incomplete, inaccurate, and in some cases even biased analysis of the South from a Northern perspective. The advocates of Southern criminology criticize the dominant logic produced in the 'modern' or 'Western' societies of the global North that social occurrences in the 'periphery' should be examined from the point of view of universal theories and laws of advancement (Connell, 2007).

Just as the raw materials of the South are taken and processed by the North and serve the interests of the North, so too, in social science research, the data of the South are handled by the North, mostly according to the theories, perspectives, and interests of the

North. Connell defines this Northern dominance over the South as 'metropolitan' thinking (Connell, 2007, p. 215). This thinking is dominant in criminology (Carrington et al., 2016). Goyes (2019a), referring to the environmental realm, calls this "ecological discrimination".

The school of Southern green criminology pays attention to the fact that environmental problems are geographically and socially uneven. There is a global mining economy that involves oil and gas and many other resources, including water. Mineral and precious metal markets generate extensive ecological damage (Lynch, 2016). Yet, little attention has been paid to those living in the sites of resource extraction—i.e. the green victimization of native/Indigenous people (de Carvalho et al., 2020; Goyes, 2019b; Goyes & South, 2021; Goyes et al., 2021). All these crimes are the result of corporations seeking out their next opportunity for exploiting the ecosystem to generate profits (Lynch, 2020, p. 58). The exploitation of the raw materials of the South by the North, which has been going on for centuries, and the fact that the North makes a high contribution to carbon emissions and does not take measures to prevent this, stands out as a very important green crime.

The Political, Economic, and Ecological Context of Turkey

Situated between Europe and Asia, Turkey is a transcontinental country, meaning that the country's borders contain land areas that are part of more than one continent. With a population of about 84 million, Turkey is both a European and an Asian country, and can be considered a Southern country. The Turkish Ministry of Trade (2019) reports that the country has the largest resources of minerals in the world.

Turkey is a country rich in biodiversity, mainly minerals. The country is attractive for investment not only for national but also for multinational corporations. Due to neoliberalism, a drive towards the depoliticization of the social and political realm through economization (Madra & Adaman, quoted in Adaman et al., 2019, p. 519) has resulted in lands and nature in cities and rural areas being open to exploitation to generate income (Adaman et al., 2019; Altıparmak, 2020; Arsel et al., 2015; Coban, 2004). Urban and rural lands, mountains, forests, rivers, seas, and many other agricultural habitats face the problems of commodification for the gains of privileged interest groups and corporations. "The AKP has intensified the transformation of the countryside, where new forms of dispossession and deagrarianisation open the way to an unprecedented extractivist drive. Together, neoliberal developmentalism and extractivism have resulted in growing social dissent" (Adaman et al., 2019, p. 514).

Prioritizing only economic growth, the AKP follows the aggressive neoliberal policies of modernization and industrialization (Aydin, 2019, p. 1). The scale of extractive industries is having destructive, social, economic, and health impacts on people. Gündüz (2015) describes this political economy model as "fetishizing growth with fatal results". The activities of international and national corporations are in dire need of scrutiny. Turkey lacks accountability; there is also an urgent need for the introduction of substantial laws against destructive extractive businesses. As part of a strategy to establish national hegemony (Adaman et al., 2019), the AKP criminalizes victims of green crime and dissenting groups. It also employs violent tactics against them.

Within the undemocratic political structure and dynamics of power in Turkey, a new form of colonialism and exploitation emerges. Extractive industries and other environmentally destructive energy projects are carried out without receiving environmental impact assessment reports and are boosted by rapid expropriation policies. Either the decisions of the administrative courts in favor of the local people are not implemented or extractive businesses continue their activities unlawfully before the court process is concluded. While the efforts of the villagers and environmentalists to stop the activities of the extractive enterprises are successful in some places, the use of violence by security guards in other places leaves the reactions of the victims inconclusive. In Turkey, all individual and collective protest actions carry social, economic, and political risks. Authorities with political power play the cards of stigmatization, blame, exclusion, criminalization, demonization, and the deprivation of economic opportunities.

Extractive businesses have significant economic, social, health, and political impacts on people. They also cause environmental devastation and victimize many, ruining the livelihoods of rural and urban people with large-scale deforestation, the reduction of agricultural and recreation areas, and biodiversity loss. Apart from the exploitation of minerals, mines, and raw materials, it was revealed that Turkey is the largest destination for waste exported from the European Union, with a volume of around 13.7 million tons in 2020 (Eurostat, 2021). While industrial, Northern countries receive raw materials for next to nothing from the South, they use these minerals in their industries. After the production process, they send the toxic waste to the South. Thus, including Turkey, the South has become the toxic dumpsite of the North (Inger, 2020; Ugurtas, 2020).

This deagrarianization policy goes back to the 1980s when the influence of international players such as the International Monetary Fund and the World Bank increased. Within the liberalization policies, agricultural subsidies were removed (Adaman et al., 2019, p. 522). Hence, people living in agricultural areas were left with no choice other than to work in adverse conditions in mines, construction places, and other industries where the life and safety of workers are at great risk. In Turkey, what Lynch et al. (quoted in Lynch, 2020, p. 6) propose is the case: given the connection between capitalism, corporate expansion, and ecological destruction/disorganization, however, green crime recognizes that laws and environmental policies, at best, limit some, but cannot solve the ecological damage caused by overproduction, overconsumption, and economic growth associated with the inherent expansionary tendencies of capitalism.

Highways; ports; airports; warehouses; dams; nuclear, thermal, and hydropower plants; factories; mine excavations; mega-tourist facilities, mega-hotels, luxury housing projects, and many other construction projects are presented by the AKP to the public as a policy of economic growth. These projects are also introduced as a solution to unemployment and as projects of modernization and industrialization, although these environmentally destructive projects only serve the interests of a small number of local and globally elite groups with which Turkish politicians ally. As a result of this policy, the lands of local people are taken from their hands with unfair expropriation policies; agricultural products are then devalued.

The adverse effects of economic development in the country have given rise to complaints against the current and potential impacts from natural resource extraction, land-use change, energy production, and increased pollution, inciting local communities at grassroots levels, as well as national and international civil society organizations, to be increasingly involved in environmental justice movements (Özkaynak, quoted in Aydin, 2019, p. 3). Over the years, the country has witnessed many well-known environmental protests such as the Bergama movement against gold extraction (Çoban, 2004; Özen & Küskü, 2009), the movement in Gerze against a coal-fired power plant (Akbulut, quoted in Arsel et al., 2015), and there are many other cases (Kap, 2020) where local communities are fighting against activities such as dam construction and energy projects in protected areas, waiving the obligatory Environmental Impact Assessment Report for mega-projects, allowing mining exploration in nature conservation areas, and weakening control mechanisms concerning the use of forest and coastal areas (Özkaynak et al., quoted in Aydin, 2019, p. 3).

Environmental activists living in cities travel long distances by bus and take part in collective action against environmental destruction in different parts of Turkey. Destinations for this collective action include mines, hydroelectric power plant (HEPP) construction, quarries in the EBS region, olive trees at risk of being felled in the Aegean Region, forests, the construction of nuclear power plants, and coastal strips. Some of the associations and platforms established for defending the river and forest areas at risk are called the 'Brotherhood of the Rivers, and the 'Brotherhood of the Forests'. The date of the first big battle against hydropower in the EBS region goes back to the 1990s (Hattam, 2015). Today, most of the valleys in the EBS region are at risk due to HEPP projects implemented or planned. Under the oppressive political climate, attempts are invariably made to neutralize the demonstrators by the violence of riot police; they are taken to court and even deprived of their freedom with baseless accusations.

The Soma Mining Disaster and the Ikizdere Stone Pitch Project

The local contractors of the multinational companies of the North who want to benefit economically by ignoring nature and human life challenge the villagers of Turkey, who defend their living spaces against extractive corporations. In one example of many, the chairman of one of these companies said: "We are patient, we wait. Keep vigil in your forests as long as you want … It takes thousands of years for a mine to form. We can wait a few more years if necessary!" (DW, 2021). What is the source of the confidence behind this arrogance, the disregard of the right to life, the normalization of environmental destruction, and challenging many more harms with a disrespectful, bossy attitude without being embarrassed, or afraid? His firm belief is that there is nothing he cannot buy for the transnational capitalist company he is a subcontractor for. Are there not any ministries and other related authorities, judicial institutions, and other official agencies in charge of protecting the environment in Turkey? Was this catastrophic green crime reduced to a conflict between an elderly man/woman who had to defend a relatively small area of land and the security forces sent by the political power defending the interests of the company? Ultimately, what are the social structures that allow large green crime in Turkey?

The answers lie with the environmentally destructive deagrarization political economy policies of the AKP. The AKP applies two different tactics. The first is trying to persuade and get the consent of the villagers whose land is at risk of environmental catastrophe and of the general public who are sensitive to the environment and local agricultural production. The second is to resort to violence against people mobilizing against environmental destruction and the landowners who want to save their livelihood and take the lands from the hands of the owners by force. Both strategies reveal the corruption and lack of democracy in Turkey. They are also present in the two case studies to which we turn now.

The Soma Mine Disaster (SMD)

Situated in the west of Turkey, with a population of about 110,000, Soma is a district of the Manisa province. Three hundred and one men lost their lives in the worst mining disaster in Turkish history at the SMD in 2014. Before SMD, 263 were killed in a 1992 accident in Zonguldak province. The catastrophe at the SMD was not a surprise. The results from technical reports, observations, and the experiences of the employees had long warned of the coming of such an appalling disaster. One of the main reasons behind the disaster was that the fire started due to the extraction of coal by a private mining company in a previously abandoned mine. The Report prepared by the Chamber of Chemical Engineers of Turkey claims that coal burning continued in the old galleries, and these areas were entered without extinguishing the fire and obtaining the necessary work permits. It was acknowledged that the carbon monoxide produced by the burning coal was a result of the collapse of the closed gallery systems due to unauthorized and uncontrolled entry into these quarries and spread throughout the mine in a short time (TMMO, 2014).

Before the SMD, the political opposition parties took some steps in order to improve safety in the mine. But the calls for improvement remained unattended because members of the governing party rejected the proposal. Works to improve the safety standards in the mine would mean increased costs to the corporations. Just as happened in the Soma mines, the strength of the AKP, and the inadequacy of social awareness, blocked the steps needed to be taken to prevent green crime fatalities. When looking at the technical reports, complaints, and parliamentary research motions about the negative conditions of the SMD, it is clear that what happened was not an accident, but murder. Thousands of miners continued to work in the mine, despite the result of the reports published before the SMD. Then, Prime Minister Erdoğan and the AKP said that the disaster was not an accident caused by negligence and/or any other reason. In a classical neutralizing move, they claimed that it was inherent in the coal mining business.

In Soma, after the mining disaster, a large mass of people protested against the AKP and Erdoğan. Another large protest was also held in

Istanbul, the most populated city in Turkey. Those who wanted to join the protests from outside were prevented from entering Soma. Angry crowds booed Erdoğan in Soma after a speech in which he said "these things happen" (*The Guardian*, 2014). Erdoğan had to shelter in a shop in the face of angry demonstrators. Security forces and some civilians used heavy violence against the protestors. Scenes of one of Erdoğan's advisers kicking a demonstrator who collapsed to the ground, and his being hit with a baton of the security forces, were caught on camera (*National Post*, 2014). The protests broke out across the country. Protesters accused the government of failing to implement safety codes at the mine. The government insisted the mine was properly checked. Erdoğan told the public not to pay attention to "extreme elements, that want to abuse developments like this one". In Istanbul, security forces barricaded Taksim Square and closed Gezi Park. The police fired tear gas and used water cannons to disperse a crowd of several thousand demonstrators who gathered to protest over the mine disaster (*The Guardian*, 2014).

The International Labor Organization ranked Turkey third worst in the world for worker deaths in 2012 (*The Guardian*, 2014). The fact that the survivors of the SMD disaster had to continue working in the mine is a matter of debate in itself. While the SMD is a story of economic, political, and social violence, it can also be considered a crisis provoked by management failure. No law was issued based on learning from the disaster. Due to both political and economic concerns, mineworkers continued their work. Adaman et al. (2019, p. 523) argue that the locals were left with no choice but to abandon their old agrarian lifestyle and start to work in mining, after a series of policies had all but destroyed the viability of peasant agriculture in the area.

The Ikizdere Stone Pitch Project

As well as in other parts of the country, biodiversity is at great risk in the EBS region of Turkey. Associations against green crime have been established in different parts of the country. These associations strive to survive and continue protesting despite all kinds of economic, social, and political obstacles. Such associations emerge as grassroots movements from different social segments and political views. The aim of being against the opening of mines and protecting nature is to defend the 'right to life'. In particular, the mines and quarries opened in different parts of Turkey will increase the risk of landslides, especially in the EBS region. Another danger is that the metals released during the storage and transportation of the material in the mine will come into contact with water and air and spread to the environment. This will adversely affect public health in addition to the damage to the environment.

The EBS region was determined by the World Conservation Foundation as one of the 200 priority areas in terms of biodiversity protection. Ikizdere Valley has also been announced by the Ministry of Culture and Tourism as a new destination area for thermal and winter tourism in the EBS region. Dozens of regions/valleys in the EBS region are at risk. For example, according to the TEMA Foundation report, 71% of Artvin province in the region has mining licenses (K2, n.d.). Ikizdere Valley is just one of the regions at risk of deterioration where the loss of biodiversity is rife and much harm is irreversible. A logistical port project was planned to be built by filling the sea with basalt in Ikizdere Valley. The basalt quarry project, which is planned to be opened in Cevizlik village in the same valley, will take about 16 million tons of stone. This project caused reactions not only from local people but also from individuals and environmentalist groups across Turkey. Local people, who defend the natural habitat of the green valley, would disappear from the area, when forced to migrate to cities upon the imminent destruction of their villages. Hence, they started a vigil with the message "We are not against investment, but against the quarry that was chosen in the wrong place, we have nowhere to go. We will not give up on our nature".

The activists said that 200,000 tons of tea grows and 1000 tons of honey is produced in the village. They also said that their product has a large share in meeting their nutritional needs. The villagers insisted that it would cost less money and environmental damage if the quarry was opened in another area closer to the planned port (Sözcü, 2021).

Despite the resistance from the opposition parties, the public, and environmental activists, the project continued with the permission of the government (Akduman, 2022). The local public took the project to the court. Although the court stated that "expecting the projects in question to be completely harmless to the environment would not be in line with the realities of life; there may be some negative consequences in terms of both humans and plants/animals", it decided to reject the lawsuit (Evrensel, 2022). The local people and the İkizdere Environment Association revealed that the river was not protected, as the court suggested, but was filled with stones by the company that carried out the project (Akduman, 2022).

The elderly did not want to leave. They wanted to remain where they have spent their lives. They had no other places to go, no homes, and no other means of livelihood. As the vegetation of the region changed, some of the previously grown vegetables and fruits were no longer available. Fertile lands were reduced. Their living spaces were taken from traditional inhabitants directly and indirectly. Those who had a house and land in the EBS region had to migrate to Istanbul and other industrial cities—without really cutting ties with their villages. They tend to spend their summer holidays in their villages and continue their agricultural activities, especially tea growing and picking. Some people after retirement in the cities return back to their villages.

The fact that the population of the region has a village–urban connection provides important opportunities for environmental awareness, organization, and collective resistance actions throughout the year. Solidarity and resistance rallies with high participation are held in big cities, especially in Istanbul. Urban and rural activists participate in rallies and other forms of resistance in the countryside. Through solidarity and joint actions, they benefit from each other's experiences and create new activist networks. These networks are also connected with international networks.

In all green crime resistance movements, participants are stigmatized as traitors and enemies of the state. Demonstrations after the SMD were stopped by very heavy police violence, and people in plain clothes attacked the aggrieved demonstrators. Thus, it was announced to the whole of Turkey that any possible collective protest actions in the future would be neutralized by the AKP, if necessary.

Conclusion

The two case studies here reviewed demonstrate Turkey's lack of political accountability. On the one hand, in securitized and sacritized politics, the SMD shows how green crime victimization is ignored by morally and financially corrupt authorities. On the other hand, the ISPP resistance movements demonstrate the ongoing efforts to resist green crime victimization and the hope for changes in the country.

The social movements protesting around the SMD and the ISPP projects are a reflection of the scale of structural irresponsibility and negligence in mass extraction projects. The two cases show the extent of green crime victimization—but also resistance—in Turkey. Due to mass extraction and power plant projects, environmental destruction has been increasing, causing the deterioration of the ecological balance and agriculture, and the disappearance of agricultural and animal husbandry areas. The villagers, whose local production areas are taken from them, are forced to leave their lands and work in extremely unfavorable conditions, deprived of economic and social security, as in mining.

Turkey has very large areas of agriculture and animal husbandry. The share of agriculture in the total production is decreasing in the country. The role of rural players has started to weaken, and villagers are threatened with "either you sell your land or it will be expropriated" (Duvar, 2021). Turkey now imports even the most basic agricultural products which used to be produced in the country. The balanced relations between rivers, fields, forests, seas, animals, crops, and people have come to an end. This causes increases in the price of agricultural products. Now, the country faces a shortage of food, even a crisis. These negative developments mostly affect women. As Shiva (1989) proposes, the productive nature of relationships with women will come to an end.

National and international corporations are the winners of the neoliberal economic policies implemented by the AKP. The losers are villagers, whose agricultural production areas are destroyed while they become a cheap labor force for corporations. Because the drivers of environmental victimization in Turkey have a global dimension, the harms described in this chapter are not unique to Turkey. These problems are experienced in many areas of the global South. As White (2018,

p. 281) argues, "transnational environmental harms and crimes are always located somewhere and need to be put into specific regional and national contexts, they simultaneously need to be analyzed in terms of international relationships, transferences, and influences".

SMD was one of the most painful events, both in terms of world mining history and the history of twenty-first-century mining accidents. As the investigation and observation report prepared by The Association for Human Rights and Solidarity for the Oppressed (2015) states, this was due to: "immoral ambition for profits, the lease system, international agreements that have not been signed until now because of increasing operating costs, unapplied and uncontrolled security measures, and a state that escapes transparency and accountability, distributing concessions". The report assigns the responsibility to the state while it questions the political pressure, economic, physical, and psychological oppression, and threats against victims of green crime. Since the SMD happened, nothing has *structurally* changed despite the criticisms and resistance by movements against environmental destruction. One of the reasons for this lack of improvements in tackling green crime victimization is that the ruling AKP government identifies environmentalism almost exclusively with the secular opposition, as opposed to treating environmental advocates as independent actors seeking to advance universalist values. The tension between environmentalists and the ruling AKP has reinforced an oppositional secularist identity among some environmental advocates, further exacerbating polarization in the environmental arena. (Özler & Obach, 2018).

For pragmatic reasons, a large section of the supporters of the AKP ignores all kinds of injustice done by the party, including green victimization. They think that if the AKP loses power, they will lose as well. For this reason, efforts to show discontent for green victimization are sidelined. They claim that participating in collective actions to resist green crime is a challenge to the state. The politics of polarization become easier. On every political issue, including debates on green victimization, the AKP polarizes the country by dividing people as "we versus them". This is evident in Soma as the results of local and national elections, held five years before the time of writing, which show that victims are forced to consent to the AKP. In local and national elections, in small local neighborhoods and/similar areas, who votes for whom is known. Democracy is as low as corruption is high.

The cases of the SMD and ISPP clearly show that Turkey has been failing at introducing the necessary institutional arrangements to benefit society as a whole. Given the current unequal political and economic structure, only privileged groups enjoy the benefits of the economy, at the expense of leaving the rest vulnerable. In terms of natural resources and the environment, the country remains open to exploitation by national and international extraction corporations. As seen in the ongoing ISPR movements, this catastrophic, economic, political, and social climate puts the people as a whole at risk of becoming victims of green crime and in confrontation with the security forces during collective action to resist such crime. Unless substantial, structural, political, and economic changes are made, green victimization will increase.

Acknowledgements I would like to thank Dr. Goyes so much for his valuable criticisms, comments, suggestions, and excellent edits on the earlier versions of the manuscript.

References

Adaman, F., Arsel, M., & Akbulut, B. (2019). Neoliberal developmentalism, authoritarian populism, and extractivism in the countryside: The Soma mining disaster in Turkey. *The Journal of Peasant Studies, 46*(3), 514–536. https://doi.org/10.1080/03066150.2018.1515737

Akduman, I. (2022, November 8). Cengiz İnşaat İkizdere'de dereye taş doldurdu. *Sözcü.* Retrieved from https://www.sozcu.com.tr/2022/gundem/cengiz-insaat-ikizderede-dereyi-tas-doldurdu-7466957/

Altıparmak, Ö. (2020, December 16). Soykırımdan ekokrıma uzanan yol. *Bianet*. Retrieved from https://m.bianet.org/bianet/iklim-krizi/236032-soykirimdan-ekokirima-uzanan-yol

Arsel, M., Akbulut, B., & Adaman, F. (2015). Environmentalism of the malcontent: Anatomy of an anti-coal power plant struggle in Turkey. *Journal of Peasant Studies, 42*(2), 371–395. https://doi.org/10.1080/03066150.2014.971766

Aydin, C. I. (2019). Identifying ecological distribution conflicts around the inter-regional flow of energy in Turkey: A Mapping exercise. *Frontiers in Energy Research, 7*, Article 33. https://doi.org/10.3389/fenrg.2019.00033

Brisman, A., & South, N. (2019). Green criminology and environmental crimes and harms. *Sociology Compass, 13*(1), e12650. https://doi.org/10.1111/soc4.12650

Carrington, K. (2021, August 11). Decolonizing criminology through the inclusion of epistemologies of the South. *The BSC Blog*. Retrieved from https://thebscblog.wordpress.com/2021/08/11/decolonizing-criminology-through-the-inclusion-of-epistemologies-of-the-south/

Carrington, K., Hogg, R., & Sozzo, M. (2016). Southern criminology. *The British Journal of Criminology, 56*(1), 1–20. https://doi.org/10.1093/bjc/azv083

Carrington, K., Hogg, R., Scott, J., Sozzo, M., & Walters, R. (2018a). *Southern criminology*. Routledge.

Carrington, K., Hogg, R., Scott, J., & Sozzo, M. (Eds.). (2018b). *The Palgrave handbook of criminology and the global south.* Palgrave Macmillan.

Coban, A. (2004). Community-based ecological resistance: The Bergama movement in Turkey. *Environmental Politics, 13*(2), 438–460. https://doi.org/10.1080/0964401042000209658

Connell, R. (2007). *Southern theory: The global dynamics of knowledge in the social sciences*. Allen and Unwin.

de Carvalho, S., Goyes, D. R., & Vegh Weis, V. (2020). Politics and indigenous victimization: The case of Brazil. *The British Journal of Criminology, 61*(1), 251–271. https://doi.org/10.1093/bjc/azaa060

Duvar. (2021, May 16). Cengiz Holding'ten köylüleri tehdit: Arazileri satmazsanız kamulaştırma isteriz. *Duvar.* Retrieved from https://www.gazeteduvar.com.tr/cengiz-holdingden-koyluye-tehdit-arazileri-satmazsaniz-bakanliktan-kamulastirma-isteriz-haber-1522427

DW. (2021). Alamos'un Türkiye İştiraki: Cevherin her hakkı 60 yıl bizde. *DW Türkçe*. Retrieved from https://www.youtube.com/watch?v=6dwwdoesEUwastego?

Eurostat. (2021, April 20). Where does EU waste go? *Eurostat*. Retrieved from https://ec.europa.eu/eurostat/en/web/products-eurostat-news/-/ddn-20210420-1?s=09

Evrensel. (2022, August 24). Mahkeme, İkizdere'deki taş ocağı projesinin durdurulması talebini reddetti. *Evrensel.* Retrieved from https://www.evrensel.net/haber/468784/mahkeme-ikizderedeki-tas-ocagi-projesinin-durdurulmasi-talebini-reddetti

FAO. (2021). *State of land, soil and water for food and agriculture–Systems at breaking point. Synthesis report 2021*. Retrieved from https://doi.org/10.4060/cb7654en

Fletcher, S. (2021). Ecocide: Kill the corporation before it kills us by D. Whyte. *State Crime Journal, 10*(2), 326–328. https://www.jstor.org/stable/10.13169/statecrime.10.2.0326#metadata_info_tab_contents

Goyes, D. R. (2019a). *Southern green criminology: A science to end ecological discrimination*. Emerald Group Publishing.

Goyes, D. R. (2019b). Cultures of exclusion: Environmental destruction and indigenous neglection. *The Critical Criminologist: Spotlight, 27*(2), 18–20.

Goyes, D. R. (2021). Environmental crime in Latin America and Southern Green Criminology. *Oxford Research Encyclopedia of Criminology*. https://doi.org/10.1093/acrefore/9780190264079.013.588

Goyes, D. R., & South, N. (2017). Green criminology before 'green criminology': Amnesia and absences. *Critical Criminology, 25*(2), 165–181. https://doi.org/10.1007/s10612-017-9357-8

Goyes, D. R., & South, N. (2021). Indigenous worlds and criminological exclusion: A call to reorientate the criminological compass. *International Journal for Crime, Justice and Social Democracy, 10*(3), 115–128. https://doi.org/10.5204/ijcjsd.1983

Goyes, D. R., Abaibira, M. A., Baicué, P., Cuchimba, A., Ñeñetofe, D. T. R., Sollund, R., South, N., & Wyatt, T. (2021). Southern green cultural criminology and environmental crime prevention: Representations of nature within four Colombian indigenous communities. *Critical Criminology, 29*(3), 469–485. https://doi.org/10.1007/s10612-021-09582-0

Gündüz, Z. Y. (2015). The "New Turkey": Fetishizing growth with fatal results. *Monthly Review, 67*(2), 40–53. Retrieved from https://monthlyreview.org/2015/06/01/the-new-turkey-fetishizing-growth-with-fatal-results/

Hattam, J. (2015, June 19). New solidarity in the struggle to protect Turkey's 'life spaces'. *Mongabay*. Retrieved from https://news.mongabay.com/2015/06/new-solidarity-in-struggle-to-protect-turkeys-life-spaces/

Higgins, P., Short, D., & South, N. (2013). Protecting the planet: A proposal for a law of ecocide. *Crime, Law and Social Change, 59*(3), 251–266. https://doi.org/10.1007/s10611-013-9413-6

Human Rights Watch. (2021). *World report 2021, events of 2020*. Retrieved from https://www.chrome-extension://efaidnbmnnnibpcajpcglclefindmkaj/; https://www.hrw.org/sites/default/files/media_2021/01/2021_hrw_world_report.pdf

Inger, A. (2020, February 13). Africa must tell the rest of the world that we are not their dumping ground. *Mail & Guardian*. Retrieved from https://mg.co.za/article/2020-02-13-africa-must-tell-the-rest-of-the-world-that-we-are-not-their-dumping-ground/

Jarrell, M. L., Michael J.L., & Paul B. S. (2014) Green criminology and green victimization. In B. A. Arrigo & H. Y. Bersot (Eds.), *The Routledge international handbook of international crime and justice studies* (pp. 423–444). Routledge. Retrieved from https://www.fondazionegermani.org/wp-content/uploads/2018/04/libri%20e%20documenti/Arrigo%20&%20Bersot%20(eds.)%20,%20The%20Routledge%20Handbook%20of%20International%20Crime%20and%20Justice%20Studies%20(Routledge,%202013)(2).pdf#page=454

Kap, D. (2020, November 16). Türkiye'de Yerel Çevre Hareketleri (1). *Sivil Sayfalar*. Retrieved from https://www.sivilsayfalar.org/2020/11/16/turkiyede-yerel-cevre-hareketleri-1/

K2. (n.d.). Artvin'i bekleyen gelecek: 'Zehirlenmiş su, yok edilmiş tarım ve ormanlar'. Retrieved from https://k2haber.com.tr/2021/01/10/artvinde-ne-orman-ne-tarim-alani-ne-de-berrak-akan-bir-su-kalacak/

Lynch, M. (1990). The greening of criminology: A perspective on the 1990s. *The Critical Criminologist, 2*(3), 3–4.

Lynch, M. J. (2016). Green criminology and social Justice: A reexamination of the Lynemouth plant closing and the political-economic causes of environmental and social injustice. *Critical Sociology, 43*(3), 449–464. https://doi.org/10.1177/0896920515598563

Lynch, M., & Stretesky, P. (2003). The meaning of green: Towards a clarification of the term green and its meaning for the development of a green criminology. *Theoretical Criminology, 7*(2), 217–238.

Lynch, M. J. (2020). Green criminology and environmental crime: Criminology that matters in the age of global ecological collapse. *Journal of White Collar and Corporate Crime, 1* (1), 50–61. https://doi.org/10.1177/2631309X19876930

National Post. (2014, May 15). Turkish prime minister says mine disasters are 'ordinary' as his aide photographed kicking a demonstrator. https://nationalpost.com/news/turkish-prime-minister-says-mine-disasters-are-ordinary-as-his-aide-photographed-kicking-a-demonstrator

Özen, Ş, & Küskü, F. (2009). Corporate environmental citizenship variation in developing countries: An institutional framework. *Journal of Business Ethics, 89*(2), 297–313. https://doi.org/10.1007/s10551-008-0001-0

Özler, Şİ, & Obach, B. (2018). Polarization and the environmental movement in Turkey. *Journal of Civil Society, 14*(4), 311–327. https://doi.org/10.1080/17448689.2018.1518773

Shiva, V. (1989). *Staying alive: Women, ecology and development*. Zed Books.

South, N. (1998). A green field for criminology? A proposal for a perspective. *Theoretical Criminology, 2*(2), 211–233. https://doi.org/10.1177/1362480698002002004

South, N. (2014). Green criminology: Reflections, connections, horizons. *International Journal for Crime, Justice and Social Democracy, 3*(2), 5–20. https://doi.org/10.5204/ijcjsd.v3i2.172

Sözcü. (2021, April 1). Köylülerden Taş Ocağı Nöbeti. *Sözcü.* Retrieved from https://www.sozcu.com.tr/2021/gundem/koylulerden-tas-ocagi-nobeti-6347208/

The Association for Human Rights and Solidarity for the Oppressed. (2015). Investigation & observation report on the Soma coal mine disaster. https://istanbul.mazlumder.org/webimage/files/soma-raporu-2015-soma-report.pdf

The Guardian. (2014, May 14). Turkey mine disaster: Protests break out after Erdogan speech. Retrieved from https://www.theguardian.com/world/2014/may/14/turkey-mine-explosion-rescue-operation-live-updates

The Republic of Turkey, Ministry of Trade. (2019). *Industry mining.* Retrieved from https://www.trade.gov.tr/data/5b8fd5bf13b8761f041fee9b/Mining.pdf

TMMO Kimya Mühendisleri Odası. (2014). *Soma Maden Kazası (!) Hakkında Kimya Mühendisleri Odası Heyetinin 14-15-16 Mayıs Tarihlerindeki İnceleme Raporu.* http://kimyamuhendisiadayi.blogspot.com/2014/05/soma-maden-ocagi-hakkinda-teknik-rapor.html

Ugurtas, S. (2020, September 18). Why Turkey became Europe's garbage dump? *Politico.* Retrieved from https://www.politico.eu/article/why-turkey-became-europes-garbage-dump/

United States Embassy and Consulates in Turkey. (2020). *2019 country reports on human rights practices: Turkey.* Retrieved from https://tr.usembassy.gov/2019-country-reports-on-human-rights-practices-turkey/

Wendling, Z. A., Emerson, J. W., Esty, D. C., Levy, M. A., de Sherbinin, A., et al., (2018). *2018 environmental performance index.* Yale Center for Environmental Law & Policy. Retrieved from https://epi.yale.edu/

White, D. (2020). *What ecocide: Kill the corporation before it kills us*. Manchester University Press.

White, R. (2018). The global context of transnational environmental crime in Asia. In K. Carrington et al. (Eds.). *The Palgrave handbook of criminology and the global south* (pp. 281–300). Springer. https://doi.org/10.1007/978-3-319-65021-0_15

White, R., & Eman, K. (2020). Green criminology, water issues, human rights and private profit. In K. Eman, G. Meško, L. Segato, & M. Migliorini (Eds.), *Water, governance, and crime issues* (pp. 3–15). Springer.

4

Environmental Exploitation and Violence Against Indigenous People in Mexico

Gabriela Gallegos Martínez, José Luis Carpio Domínguez, and Jesús Ignacio Castro Salazar

Introduction

In 2020, in the Zapotec region of Loxicha in the state of Oaxaca Mexico, the young 21-year-old biologist Eugui Roy Martínez Pérez was shot to death. Eugui was one of the most outstanding herpetologists and conservationists in this region. He became a link between science and the cultural and religious traditions of his community. This region of

G. G. Martínez (✉)
Universidad de Monterrey, San Pedro Garza García, México
e-mail: gabriela.gallegos@udem.edu

J. L. C. Domínguez
Universidad Autónoma de Tamaulipas, Reynosa, México
e-mail: jcarpio@docentes.uat.edu.mx

J. I. C. Salazar
Instituto Tecnológico Superior de Abasolo, Abasolo, México
e-mail: ignacio.cs@abasolo.tecnm.mx; jicastro13@hotmail.com

D. R. Goyes (ed.), *Green Crime in the Global South*, Palgrave Studies in Green Criminology, https://doi.org/10.1007/978-3-031-27754-2_4

Oaxaca, inhabited mainly by Indigenous communities, has been characterised in recent decades by political problems, guerrilla presence, and homicides and chieftainships that go unnoticed by the rest of the country. But Eugui's case resonated nationally and internationally due to his young age, his positive influence within his community, and his relationship with the Mexican academic community.

> The door of Eugui's house, which was the scene of the crime, was not cordoned off during the first hours. An anonymous source said that the Public Ministry appeared until after noon and did not do any diligence. At 4:00 p.m., Eugui's body was still in a patrol boat. The conditions were right for the crime to be lost in Loxicha's silence, but the family asked that the investigation case be removed from the municipality and, with the media pressure generated by the case, the Oaxacan government promised to investigate. (Soberanes, 2020)

However, the murder of Eugui remains unsolved, like many deaths of Indigenous conservationists. The short-term Mexican memory of violence has resulted in deaths like Eugui´s and other Indigenous conservationists remaining unsolved and unpunished—a phenomenon co-produced by the large number of causes and aggressors (drug trafficking, private companies, government, mines, criminal groups, among others).

Traditionally, Indigenous communities have been one of the most excluded social groups in Mexico, causing the unfulfilled needs and the levels of violence they experience to remain invisible. Indigenous communities in Mexico have historically faced various types of violence: from the Spanish colonisation in the late fifteenth century (Bueno, 2016; Goyes, 2021; Goyes et al., 2021), displacement from their lands (Cárdenas, 2013; Velázquez, 2017; World Bank, 2020), social persecution for their cultural and religious beliefs (Anguiano, 2008; Collier & Rus, 2002), discrimination for their ethnic characteristics (Czarny, 2012; Durin, 2007; Gallegos-Martínez & Cruz-Salazar, 2021; Moreno, 2010), to the violation of their rights by the state (Comisión Nacional de Derechos Humanos [National Commission for Human Rights, CNDH], 2020; De Carvalho, 2021; Romero-Zepeda & Ortega-Marín, 2017).

Indigenous communities are social groups with a strong relationship to, and beliefs in, worldviews and a culture about their territories and natural environment (Goyes & South, 2021). It is estimated that globally the Indigenous population is 476 million people distributed in more than 90 countries, representing 6% of the planet's total population (World Bank, 2020). In terms of environmental conservation, Indigenous peoples have ancestral knowledge and experience in adapting to, mitigating, and reducing risks from climate change and natural disasters (World Bank, 2020). The most recent estimates state that Indigenous communities protect 80% of the planet's remaining biodiversity (World Bank, 2020).

According to data from the Instituto Nacional de Estadística y Geografía de México (INEGI; Mexican National Institute of Statistics and Geography), the population concentration of Indigenous communities in Mexico has decreased from 16% in 1930 to 6% in 2020 (INEGI, 2020a), representing a loss of cultural diversity and ancestral knowledge. This decline is related to different aspects (migratory processes, government neglect) (Gallegos-Martínez & Cruz-Salazar, 2021), but mainly and majorly due to violence against these communities.

Mexico has approximately 138 million hectares of forest vegetation on its territory, which corresponds to just over 70% of the national land area (Gobierno de México, 2014). This forest vegetation is home to most of the country's biodiversity and all the zones classified as Natural Protected Areas (NPAs). The forest area used to be mainly under communal property status, that is, it belonged to common lands and communities, including Indigenous communities. Approximately 55% of the forest area in Mexico is under communal ownership, specifically in forest and jungle areas, where common lands and communities comprise up to 60% of it (Madrid et al., 2009).

Most of Mexico's forest areas are owned and inhabited by Indigenous communities. According to estimates, of the 68 Indigenous peoples in Mexico (Gobierno de México, 2020), at least 29 live in rainforests and 28 in mesophilic mountain forests (Comisión Nacional Forestal[CONAFOR], 2013). Of the almost 19 million hectares of federal NPA in the national territory as of 2008, roughly 10% are Indigenous peoples' territories (Boege, 2008), an extent which is twice the size in area of Portugal or Hungary.

According to the Instituto Nacional para las Personas Indígenas ([National Institute for Indigenous People] INPI, 2020) and the INEGI (2020a), Mexico's Indigenous groups mainly settle in the south, east, and south-east of the national territory in the states of Oaxaca, Chiapas, Veracruz, Puebla, and Yucatán. These five states concentrate 61% of the total Indigenous-speaking population. The rest of the population concentrates in the central and northern regions of the country (38.9%) (Fig. 4.1).

The International Tropical Timber Organization estimates that between 12 and 13 million people live in forest territories in Mexico. Out of these, around five million are Indigenous people living in conditions of poverty (Food and Agriculture Organization [FAO], 2011), government neglect, and violence (Carpio-Domínguez, 2021; González-Montes, 2009; Sainz, 2018). Their recent fight to defend their territories

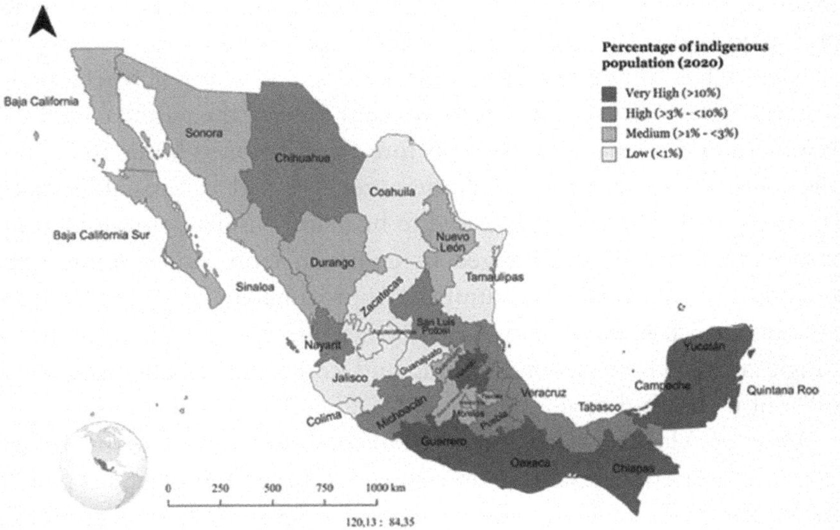

Fig. 4.1 Indigenous population concentration in each state, 2020 (*Source* Authors with supporting background information from INEGI [2020b])

and environments is in response to large industrial projects that cause displacement and to violence by criminal drug trafficking groups, which results in severe environmental problems.

Conflict Over Indigenous Territories

The concept of *territory* is key to understanding Indigenous peoples. The conceptualisation of the term 'territory' in the Indigenous context has several issues. Territorial systems are not only political spaces (geographically delimited by borders, states, municipalities, laws, etc.). *Territory* for Indigenous peoples means also a space in which diverse cultural and traditional idiosyncrasies about the meaning of nature and its conservation converge. Several studies in Mexico have focused on analysing socio-territorial disputes between Indigenous groups and government or private industries (FAO, 2011; Galván-Meza et al., 2019; Martínez-Coria & Haro-Encinas, 2015; Rodríguez-Wallenius, 2017) in which it is reported that public and private institutions take advantage of and create legal loopholes in the legal definitions of 'territory' (Boege, 2008; Haesbaert, 2011) to exclude, dispossess, and evict Indigenous communities from the places they have historically inhabited.

Concerning the rights of Indigenous peoples, Mexico has an institutional structure that, in theory, monitors, protects, and defends Indigenous rights. The structure is based on a legal framework that includes the Mexican Political Constitution, the Comisión Nacional de Derechos Humanos (National Commission for Human Rights), and the Convention 169 on Indigenous and Tribal Peoples of the International Labour Organization (ILO), and the Agrarian Law and the General Law on Sustainable Forestry Development and its regulations. This framework considers that the use of natural resources, the management of the territory, and environmental conservation are interrelated to Indigenous communities.

In practice, however, the legal framework contributes to the expropriation of Indigenous territories. (Telling is that the laws have not been translated into Indigenous languages.) For example, Article 2 of the Mexican Political Constitution recognises and guarantees the right of Indigenous peoples to self-determination and autonomy so as to access the preferential use and enjoyment of the natural resources of the places they inhabit. This same article, nevertheless, also stipulates that this does not apply to resources that are located in strategic areas and other forms and modalities of land ownership and tenure established in the Mexican Political Constitution. Article 15 of the 169 Convention establishes that if the state has rights over existing resources in the land or subsoil, the concerned Indigenous people must be consulted to ascertain their interest before authorising any exploitation project. López Bárcenas (2002) explains that such legislation takes away the power of Indigenous communities to make decisions about their territories, being limited to only inhabiting them but remaining powerless to prohibit their exploitation. Article 2 of the Mexican Political Constitution establishes a difference between strategic and non-strategic resources for the nation. Strategic resources are the property of the Mexican state, which, as stated in Article 2, Convention 169, and as explained by López Bárcenas (2002), must previously have made an agreement with the Indigenous peoples about the conditions for exploitation. This situation contributes to conflicts between Indigenous peoples, the government, and possibly private individuals interested in exploiting the biodiversity in Indigenous territories, leading to human rights violations, homicides, evictions, and intimidations against Indigenous communities.

Violence against Indigenous groups stems from outsider visions of their territories as economic-productive resources for the government and private actors. The Mexican state has made use of legal procedures that allow the dispossession of Indigenous communities of their lands (Martínez-Coria & Haro-Encinas, 2015). This process mainly benefits the development of industrial projects, both public and private, located in territories that, most of the time, belong to Indigenous groups, thus placing them in situations of risk and vulnerability (Global Witness, 2020).

The Centro Mexicano de Derecho Ambiental (Mexican Centre of Environmental Law, CEMDA) has recorded aggressions against conservationists in the country in two reports (2020 and 2021). However, it does not differentiate between the social groups victimised (Indigenous, mestizos, foreigners, etc.). So, a precise report of aggressions against

Indigenous conservationists is still pending because they face challenges that are unique to them (García & Horbath, 2019; Ortiz-Hernández, et al., 2018; Solís, 2017; Vázquez-Parra & Campos-Rivas, 2016). Importantly, Indigenous communities are preponderantly neglected when requesting protection from government institutions (CEMDA, 2020), the complaints they make to the official security institutions are not addressed, and the aggression and violations of their rights are less documented and investigated (CEMDA, 2020, 2021). Furthermore, the records of criminal statistics in Mexico do not differentiate between the causes of aggression against Indigenous conservationists. For instance, there is no differentiation between deaths due to drug trafficking and deaths due to defending territory and the environment.

Indigenous and the Global South Perspective

Problems faced by Mexican Indigenous populations (as in other parts of the world) stem from a complex socio-historical organisation underpinned by a social division that considers specific knowledge, beliefs, or practices more valid than others, that is, there is a cultural asymmetry that stigmatises, discriminates against, and makes Indigenous people invisible (Goyes, 2019). In this sense, De Sousa-Santos (2010) argues that social forms of hierarchy are implicit in the divisions among people, including the capitalist productivism monoculture and the monoculture of knowledge that assumes that only certain types of knowledge and forms of production are valid. For example, undervaluing the knowledge of Indigenous communities about herbal medicine or about agricultural production techniques prioritizes a large-scale monoculture, which also involves the use of agrochemicals, deforestation, and pollution.

Thus, underlying all forms of a monoculture of thought is a power relationship of domination–submission where specific cultural, social, political, and even physical characteristics are considered superior to others (Dietz, 2017). In this unequal relationship, the countries of the global North hold greater economic, political, and social power than those of the global South, where most Indigenous communities are located (Goyes, 2021). Wallerstein (2004) makes a similar point when he talks about the centre or periphery countries. *Centre* countries have control over economic and political institutions, technologies, and communications, while the *peripheral* countries are in charge of supplying labour and natural resources to core countries. The latter, having economic and political control, dictate the regulations, laws, or rules, making it difficult for peripheral countries to disengage from the control exercised (Goyes, 2017; Wallerstein, 2004).

Escobar (1998) and Esteva (1996) argue that the definition of 'developed' and 'developing' countries comes from a mere Western vision containing a bias in the parameters used to define and measure development. The problem with such a definition is that 'developed' countries become saviours and, as such, have power and control over the supposed countries in need of help. The so-called developed or centre countries influence, as mentioned, political and economic decisions, so they also regulate the forms of production in all regions, thus affecting 'underdeveloped' countries (Foladori, 2001). As an example of the above are the extractivist practices of mining, agriculture, and logging industries that excessively remove natural resources without taking into account the impact on the environment because the concessions under which the industries are protected are backed by international legal agreements that allow them to penetrate almost without restriction into the territories of periphery or underdeveloped countries (Pérez, 2015). Goyes (2021) points out that the environmental degradation caused by Northern practices of exploitation over the global South has not only led to crimes related to the establishment of illegal markets or the emergence of monopolies but has also exacerbated social conflicts. Industries, through polluting practices, expose Indigenous people to severe health damage and to other types of systematic violence that affect their integrity.

Indigenous peoples have been traditionally excluded from environmental and political discourses and face serious socio-environmental problems derived from the over-exploitation of nature in their territories (Toledo et al., 2015). In addition, they lack formal recognition of their lands, territories, and natural resources. They are often the last to receive public investment, access to justice, infrastructure, and participate in social representation and decision-making processes (World Bank,

2020). Indigenous peoples have become marginalised from economic, political, social, and cultural development (CNDH, 2020).

Southern green criminology in its approaches to socio-environmental conflict and crimes considers the intertwined colonial practices, monocultural thinking, and divisions of power among the global North and South (Goyes, 2019). From this perspective, it is essential to understand the diversity of roles and characteristics of the *actors* and *stakeholders* involved in the dynamics of environmental conflict. Therefore, in this chapter, we examine the Indigenous groups to which most conservationists belong. We also analyse the types of violence addressed towards them and the perpetrators' belonging.

Violence Against Indigenous Conservationists

An Approximation of the Characteristics and Conditions of Violence Against Indigenous People

From the perspective of Southern green criminology, we analyse the diversity of ethnic groups that have been victims of violence by different legal and illegal economic sectors in the country. The analysed data for this study were gathered from newspapers published nationally during the period 2010–2020 on aggression against members of Indigenous communities involved in environmental conservation activities in Mexico. We consulted 68 newspapers' digital portals: 17 national, 19 regional, and 32 local. Nine variables were analysed: year, gender, age, state, ethnic group, conservation activities, aggressors, economic sector, and typology of violence. These variables guided the consultation of the information in each of the entries. In turn, sub-variables obtained during the study allowed the data to be synthesised for the analysis.

The keywords we used on the search engines were: aggressions, Indigenous groups, conservation, community leaders, Indigenous violence, Indigenous criminalisation, megaprojects, drug trafficking and Indigenous people, mining, illegal detentions, murders of Indigenous people, deprivation of liberty of Indigenous people. The acquired data were analysed through the IBM Statistical Package for the Social Sciences (SPSS, v. 25), applying descriptive statistics for the information from 72 newspaper articles identified in the study period, and with which the saturation of the sample was achieved, since the information was replicated from national newspapers to state and local newspapers.

The statistical analysis refers to Indigenous communities by their self-designated names in their languages and not by those given in non-Indigenous languages. This practice is a form of recognition of their ethnic, cultural, and political identity. It is also a fundamental right of Indigenous populations that legitimises cultural diversity and is endorsed by different international declarations and conventions (Aguilar-Cavallo, 2006). Using the Spanish name, although more common, can also constitute a form of identity imposition, an asymmetrical practice of power and a form of non-recognition that hides cultural diversity (Bonfil-Batalla, 1972). Therefore, self-designation names are used as visualisation, revindication, and reciprocal recognition (Pérez-Ruiz, 2019).

Aggressions Against Indigenous Conservationists in Mexico During the 2010–2020 Period

CEMDA reports 564 cases of aggression against environmental and Indigenous rights conservationists in Mexico in 2012–2020 (CEMDA, 2021). As mentioned above, these data do not specify how many belonged to Indigenous communities, so we extrapolated the frequency data identified in this study with those reported by CEMDA to generate a framework for discussion of the data analysed.

Aggressions against Indigenous conservationists identified in newspaper reports in Mexico had a prevalence of 94 cases in the period 2010–2020, with an average of 8.5 cases per year. It is necessary to consider that CEMDA does not include in its studies the frequencies of aggressions in 2010 and 2011. Nonetheless, it was identified in the present study that during 2010 there were two cases of aggression against Indigenous conservationists, representing 2.1% of the national total in the study period, while in 2011, the figure increased to four cases (4.3%) (Table 4.1).

Aggressions against Indigenous conservationists in the period of analysis show an increasing trend, as illustrated in Table 4.1. Two above-average increases in cases of violence against Indigenous people (~8.5) stand out, the first during 2012 ($n = 14$, 14.9%) and the second corresponds to 2018 ($n = 15$, 16.0%) and 2019 ($n = 16$, 17.0%) (see Fig. 4.2).

The increase in cases during 2012 coincides with the Indigenous mobilisation against the implementation of the national Energy Reform. The Energy Reform entailed the development of mega-projects such as gas pipelines, petroleum exploitation, renewable energy fields, and geothermal plants. During 2012, the primary aggressor of Indigenous community members was the energy sector, as a consequence of their resistance. Of the 14 cases of violent attacks, 42% ($n = 6$) corresponded to the development of large electricity projects in the states of Guerrero, Veracruz, and Oaxaca. In addition, the changes made to the Mexican Political Constitution in 2013 in regards to the Reform allowed private

Table 4.1 Frequency of aggressions on Indigenous conservationists in Mexico during 2010–2020

Year	Conservationists[a] (*n*)	Indigenous conservationists[b] (*n*)	Percentage[b]	Percentage of aggressions on Indigenous people
2010	—	2	2.1	—
2011	—	4	4.3	—
2012	24	14	14.9	58.3
2013	64	7	7.4	10.9
2014	78	5	5.3	6.4
2015	107	6	6.4	5.6
2016	85	4	4.3	4.7
2017	53	10	10.6	18.9
2018	49	15	16.0	30.6
2019	39	16	17.0	41.0
2020	65	11	11.7	16.9
Total	564	94	100	

[a]Total reported by CEMDA (2020, 2021)
[b]Frequency and percentage based on hemerographic reports

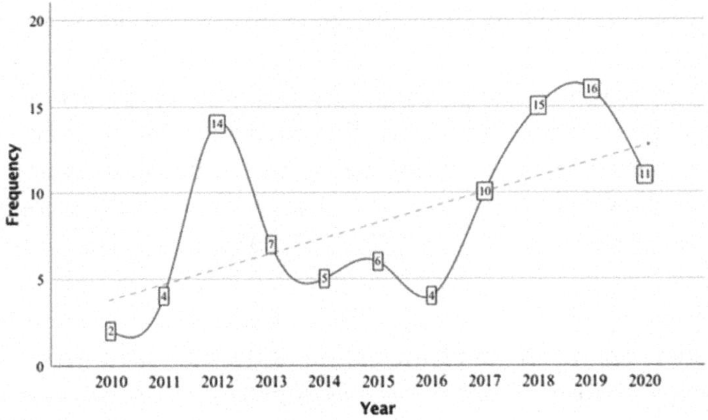

Fig. 4.2 Frequency of aggressions on Indigenous conservationists in Mexico during the period 2010–2020

companies to receive contracts to exploit energy resources. The changes were made without considering the needs or consent of the Indigenous communities that have historically inhabited the target territories (CEMDA, 2020; Rousseau, 2019).

The Energy Reform was approved in 2014 and aimed to develop the petroleum industry, a national electricity system based on technical and economic principles, and to incorporate private companies in these economic activities under the stewardship of the Mexican State (Cardoza-Zúñiga, 2017; Gobierno de México, 2013; Ruiz-Rincón, 2015). These 'development' projects were a threat to the natural, cultural, and traditional components of the Indigenous communities and thus spurred their resistance. We understand 'social resistance' as the collective action whose purpose is to minimise or eradicate a problem and implies that there is a level of organisation to define interests, activities, and even a collective identity (Melucci, 1999). In Mexico, Indigenous peoples have developed ways of resisting mega-extractivist projects. Toledo et al. (2014) categorise Indigenous resistance into two types: (a) resistance that has a defensive approach and seeks to prevent certain extractivist projects continuing to operate in their territories, and (b) resistance that is dedicated to carrying out and promoting alternative projects for the production and use of natural resources.

Some actions related to the first type of resistance are the creation of or affiliation to networks of organisations whose purposes are similar: defending against mining, agribusiness, excessive logging, or defending water (Cherem, 2016). Through these networks, Indigenous peoples articulate their efforts with those of other organisations, in legal matters, pronouncements in the media, and other acts that make the reality of environmental defenders visible, such as parades or caravans, as a way of pressuring and demanding that the government respond to them (Cherem, 2016; Valladares-De la Cruz, 2017). The second type of resistance involves "the countercurrent creation of alternative modes of articulation with nature and new ways of producing, circulating, transforming and consuming" (Toledo et al., 2014, p. 121). As an example, there are cooperatives of organic coffee growers in Chiapas or beekeepers in Oaxaca and Veracruz, whose work is based on respect for nature, solidarity, and fair trade. However, the lack of response from the government to Indigenous peoples' protestations, and the increase in violent acts against these populations, has forced them to activate other types of strategies that do not necessarily have a peaceful connotation. For instance, self-defence groups have emerged in states such as Michoacán or Guerrero (Cherem, 2016).

As Fig. 4.2 shows, the increase in the frequency of cases of violent aggression during the implementation of the Energy Reform (2013–2014) evidenced that Indigenous people's fight for their territories was criminalised as "anti-progressive" and "anti-development" (Coalición para los Derechos Humanos en el Desarrollo [Coalition for Human Rights in Development, CDHD], 2020; Rodríguez, 2002). Indigenous conservationists and environmental organisations face governmental and media stigmatisation through labels such as "anti-development", "criminals", or "terrorists" (Global Witness, 2020). Such stigmatisation is a severe problem in protecting Indigenous territories because it justifies the Mexican state's militarisation, resulting in significant impact on the protection of their environment and of their human rights (Comisión Económica para América Latina y el Caribe [CEPAL], 2014).

During the period 2018–2019, the implementation of mining activities, wind power developments, and drug trafficking resulted in another wave of aggression against Indigenous conservationists. In fact, of the entire period analysed (2000–2010) the highest frequency of violent attacks was identified in those two years. However, it is necessary to clarify that these attacks were a consequence of the Energy Reform implemented in 2014 and the conditions of insecurity derived from the activities of organised crime in the national forest territories (Carpio-Domínguez, 2021).

The frequency of cases of aggression against Indigenous people decreased during 2020. This corresponds to the first year of the COVID-19 pandemic. Although the increase in cases appears to be interrupted, the trend continues to rise with a frequency of 11 aggressions on Indigenous leaders during 2021. CEMDA (2021) reported that during 2020, 90 aggressions against conservationists were identified in Mexico, "positioning it as the most violent year for the defence of environmental human rights practice since 2018" (p. 18). When considering the general data reported by CEMDA, during the first year of the COVID-19 pandemic, 12.2% of the aggressions against conservationists were against Indigenous persons.

Various studies on conservation actions have offered a gender perspective (Global Witness, 2020; Hernández-Moreno & Nuñez-Vera, 2014; Martínez-Corona, 2003; Reyes-Aguilar, 2017; Vázquez-García, 2003; Villavicencio-Enríquez, 2011). We too acknowledge the value of considering gender, not only to establish a quantitative differentiation, but to explore the participation of Indigenous women and men in environmental conservation in violent contexts in Mexico. In the period of our analysis, we identified that 88.1% ($n = 84$) of the target of aggression against Indigenous people have been men, while women only had a percentage frequency of 11.9% ($n = 10$). That these women are reportedly less attacked during environmentalist action does not mean that they contribute less to the struggle. Rather, women lead Indigenous rights dialogues and protection, achieving significant outcomes nationally and internationally. While we know that, globally, more than one in ten defenders killed are women (Global Witness, 2020), and that women's circumstances are exacerbated when other household members are defenders, making them targets of violent aggression (Global Witness, 2020), we need more analysis for why Indigenous men are victims of direct aggression more often than women.

In Mexico, aggressions on female conservationists exceeded 18.5% in 2020. These recorded aggressions are related to gender-based violence and lack of governmental measures to "guarantee protection for the integrity of women in private or public spaces, which accentuates and reinforces the gender-based violence of which environmental and territorial women defenders have been victims" (CEMDA, 2021, p. 14). Women activists for territories and the environment need to be assuredly supported and protected (Global Witness, 2020), as suggested by national and international conventions and collaborations to which Mexico is a part of. For example, the Convention on Biological Diversity mandates that "complete, effective and inclusive participation, particularly of women in decision-making on natural resources in Indigenous peoples' territories must be ensured for them" (2018, p. 59).

The analysis of the types of violence against Indigenous people shows that the main form is homicide (n = 79, 84.0%), followed by injuries (n = 8, 8.5%), illegal detention (n = 6, 6.4%), and death threats (n = 1, 1.1%) (see Fig. 4.3). The main modus operandi in the homicides is gunshot attacks. From a criminological point of view, this means that the aggressions against this group are direct and are considered violent deaths, and consequently, crimes of high social impact. In the male population the types of aggressions are diversified, representing 75.5% (n = 71) homicides, 6.3% (n = 6) illegal detentions, 6.3% (n = 6) physical aggressions (injuries), and 1.0% (n = 1) death threats, while 8.5% (n = 8) of Indigenous women were murdered and 2.1% (n = 2) were physically assaulted (injuries).

However, some crimes, such as death threats, are not always reported. Therefore, they are not registered by government institutions, making them difficult to include in official statistics, in what is known as the black figure of crime (Sozzo, 2003). It is reasonable to think that crimes as death threats against Indigenous conservationists may be higher than reported in this study. This phenomenon has a profound impact on the execution of conservation activities. This anti-social behaviour deters the actions and efforts of Indigenous people's groups to defend their rights and territories.

Data show that aggression against Indigenous community members is concentrated in 31.2% (n = 10) of Mexico's states. The frequency

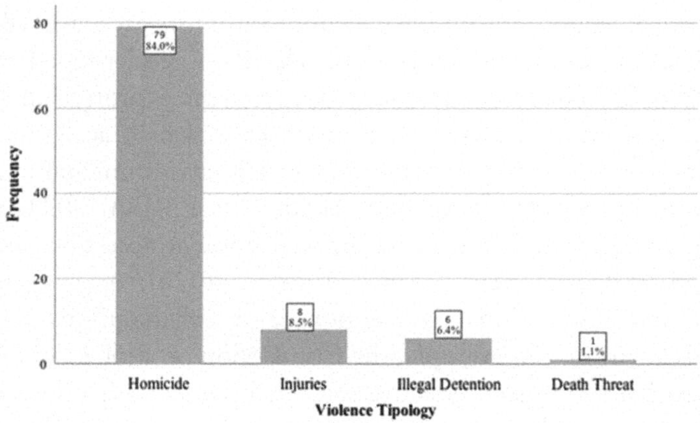

Fig. 4.3 Types of aggression against Indigenous conservationists in Mexico from 2010 to 2020

of aggressions by regional distribution corresponds to Chihuahua and Baja California in the north of the country; Nayarit, Michoacán, Puebla, Morelos, and Veracruz in the centre; and the states of Guerrero, Oaxaca, and Chiapas in the south (see Fig. 4.4).

The south of Mexico is where aggressions against Indigenous conservationists are concentrated. Oaxaca (n = 46, 48.9%) has the highest prevalence of aggression cases, followed by Guerrero (n = 18, 19.1%), while in the north the stateof Chihuahua (n = 15, 16.0%) has the highest prevalence; these states have a prevalence of >10 cases in the study period. Hence, these date are consistent with the distribution of the Indigenous population, as shown in Fig. 4.1, being concentrated in the states of Oaxaca, Chiapas, Veracruz, Puebla, and Yucatán (INEGI, 2020a, 2020b).

The frequency of aggressions against Indigenous community members increased, both in the reported cases and in the states in which were reported. That is, during the first years of analysis, the attacks occurred mainly in the southern region of the country, in Oaxaca, Guerrero, and Veracruz; in 2019 and 2020, new cases were reported in the northern region of the country, in the states of Chihuahua and Baja California Norte. Therefore, the data show a change in the geographical distribution

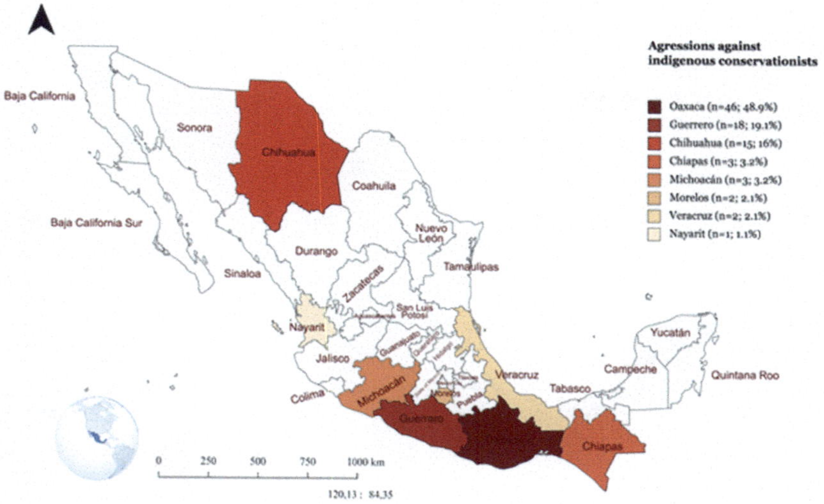

Fig. 4.4 Frequency of aggressions against Indigenous conservationists by state from 2010 to 2020 in Mexico

of aggressions against Indigenous peoples throughout the study period (see Fig. 4.5).

The identified aggressions concentrated in 13 Indigenous groups are the Binnizá group, whose inhabitants are in Oaxaca. They have suffered the highest number of aggressions (n = 31, 33.0%), followed by the Nahua people (n = 19, 20.2%) from Hidalgo, Veracruz, and Guerrero, the latter state having the second-highest rate. The third highest frequency of aggressions was against the Rarámuris (n = 15, 16.0%) in Chihuahua. In the Tachihuiin, Wixárika, Mé'Pháá, Ayuuk, Bats'i k'op, and Bats'il k'op Indigenous groups, only one case per group was identified (see Fig. 4.6).

The identified aggressions are related to the group's geographical location and the exploitable resources of each territory. For example, the Binnizá are the Indigenous group with the highest frequency of aggressions because Oaxaca has more significant heterogeneity of aggressive economic sectors (electric energy projects, mining, drug trafficking, government, and water conflicts) (see Fig. 4.7). The Nahuas inhabit

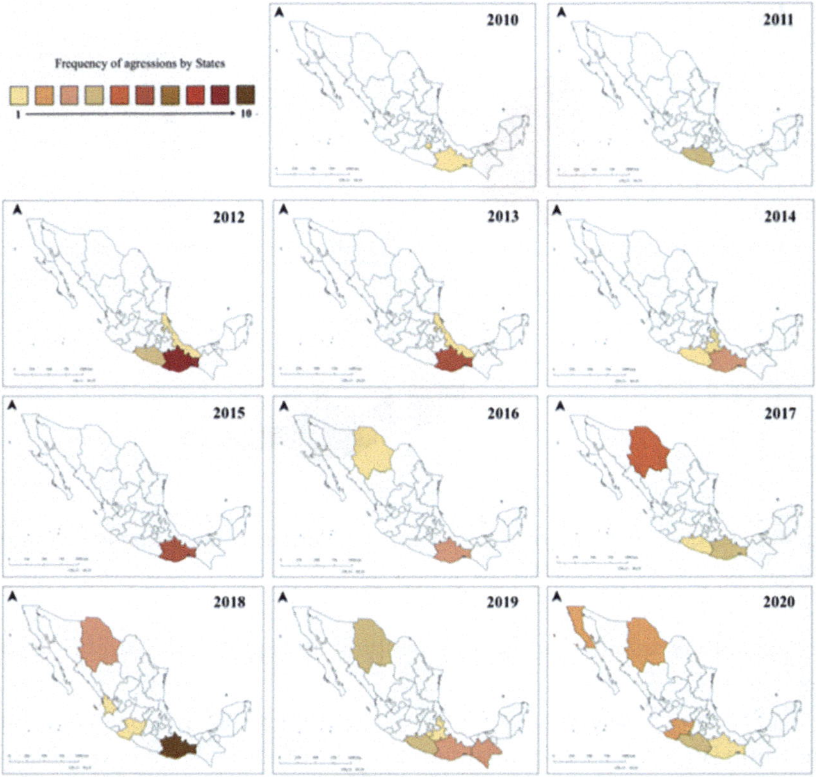

Fig. 4.5 Frequency of aggressions on Indigenous conservationists by state during the period 2010–2020 in Mexico

mainly the state of Guerrero, the state with the second most aggressive economic sectors (forestry, electricity, mining, drug trafficking, and water conflicts). Finally, the Rarámuris in the state of Chihuahua, suffer mainly with aggressions from forestry, mining, and the drug trafficking sectors.

Indigenous groups are vulnerable to sectors with more significant economic capacity, political power, and security. The primary economic sector, which is clearly identified as responsible for the most victimisation of Indigenous peoples in Mexico is drug trafficking, with 22.3% (*n* = 21) of the cases reported. This salience is related to the fact that, as an illegal activity, drug traffickers do not try to hide their identity nor the

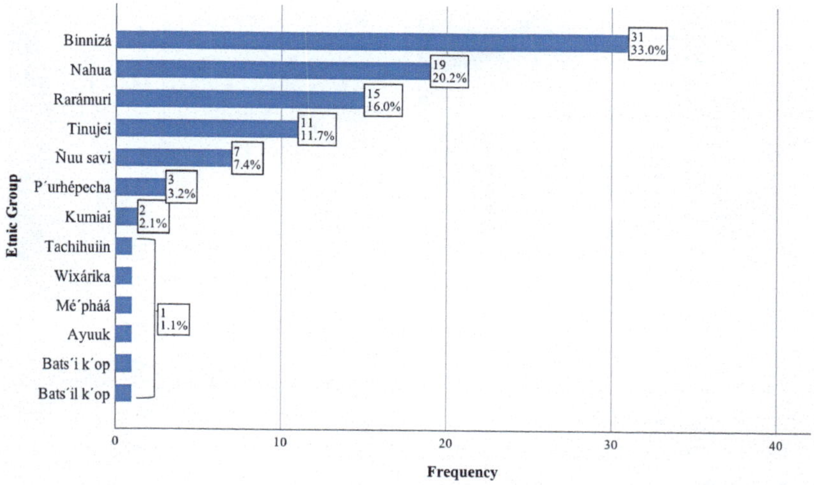

Fig. 4.6 Main Indigenous conservation groups attacked in the period 2010–2020 (*Note* Indigenous groups with a frequency of cases of aggression of 1% corresponds to one case [$n = 1$] per group)

violence. Rather, such violence is useful to them in building a reputation as a powerful sector. Second place is represented by unidentified sectors, the aggressors being unknown, and may be related to cases of intimidation to prevent those violent events that have been reported in the sectors of water management, mining, forestry, and in electrical energy projects are reported to security agencies (Fig. 4.8).

Among legal economic sectors, electrical energy projects ($n = 17$, 18.1%) and mining ($n = 15$, 16.0%) stand out because the aggressions are characterized by being carried out by private companies, by members of criminal and paramilitary groups hired by companies to intimidate and attack Indigenous people, and by the government justifying the use of police force for the repression brought about as a result of the criminalisation of Indigenous social movements (CDHD, 2020; Tetreault et al., 2012) (see Fig. 4.9). The forestry ($n = 6$, 6.4%) and water management ($n = 4$, 4.3%) sectors had the lowest frequency of aggressions, conducted by paramilitary groups, drug traffickers, and private companies. As shown in Fig. 4.8, mining is the sector with the most

Fig. 4.7 Protests in Mexico City against the forced disappearance and homicides of Indigenous people from Oaxaca (*Note* Images taken in 2020)

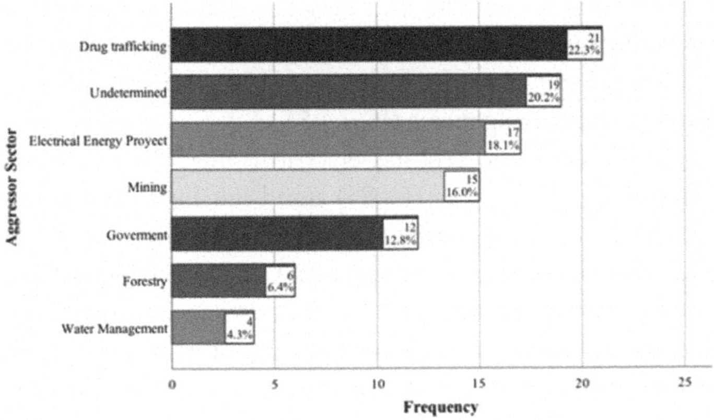

Fig. 4.8 Aggressors of Indigenous conservationists in Mexico by economic sector in the period 2010–2020

diversity of direct aggressors against Indigenous people, followed by electrical energy projects. It is documented that mining is a severe problem for Indigenous communities; various studies have reported 113 socio-environmental conflicts over mining in Mexico, of which 28 have been reported in Indigenous territories (Pérez-Jiménez, 2014; Valladares-De la Cruz, 2017).

It is reported that for 59.6% of Indigenous peoples, their ancestral territories have been granted to foreign mining companies, mainly from Canada, with severe cases of human and Indigenous rights violations (Boege, 2020; Valladares-De la Cruz, 2017). However, drug trafficking is the primary known aggressor sector of Indigenous conservation groups. Although it is not excluded from being involved in the forestry sector, its main activity is planting and harvesting narcotics in Indigenous territories (Carpio-Domínguez, 2021). Additionally, drug traffickers aid mining companies by intimidating environmental defenders who resist the predatory activities of these companies. (This criminal symbiosis is also reported in Colombia by Ocampo-Rodríguez, 2015.)

•

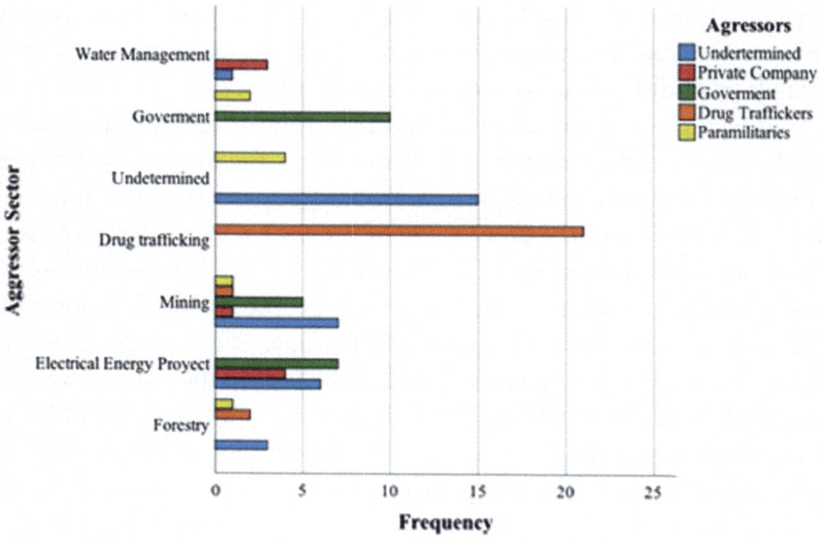

Fig. 4.9 Relationship between aggressors and economic sector in violence against Indigenous conservationists in Mexico during 2010–2020

The Need for a Southern Green Criminology Perspective in México

There is a lack of attention from the Mexican state to the protection of the rights of Indigenous communities and their territories. Hence, Indigenous peoples are forced to seek support in international contexts. This circumstance creates an "international resonance bell that allows them to vent their problems and expose the scandalous conditions of survival and the flagrant human rights violations they suffer in the hands of the social sectors with more power" (Rodríguez, 2002, p. 6). Furthermore, there is minimum state intervention focused on supporting Indigenous communities in the surveillance of forest territories in Mexico (Castro-Salazar & Luyando-Cuevas, 2020).

The living conditions of Indigenous communities in Mexico make them vulnerable. They face risks and social problems arising from inaccessibility to essential services and extractivist practices that threaten the conservation of their territories and their natural components, culture, and traditions. These problems are influenced by historical processes of invisibilisation and struggle (Paz-Salinas, 2011), which exacerbate the violence suffered by Indigenous people. Lack of recognition makes it difficult to punish crimes committed against them.

Southern green criminology is committed to a decolonising process, which recognises the colonial logic that informs the interaction between Indigenous and non-Indigenous communities, prioritises the approach to the problems from the experiences of those who have been victims of crimes, and uses their knowledge (Goyes, 2021). We have identified the need to study the situation of Indigenous people in Mexico along the lines suggested by Southern green criminology. We consider that this perspective allows us to recognise and make visible the cultural importance and implications of traditional knowledge and practices and how these violent aggressions can lead to the loss of important knowledge for the conservation of life on the planet.

References

Aguilar-Cavallo, G. (2006). La aspiración indígena a la propia identidad. *Universum, 21*(1), 106–119. https://doi.org/10.4067/S0718-23762006000100007

Anguiano, M. (2008). Chiapas: territorio de inmigración, emigración y tránsito migratorio. *Papeles de población, 14*(56), 215–232. http://www.scielo.org.mx/scielo.php?script=sci_arttext&pid=S1405-74252008000200010&lng=es&tlng=es

Boege, E. (2008). *El patrimonio biocultural de los pueblos indígenas de México. Hacia la conservación in situ de la biodiversidad y agrodiversidad en los territorios indígenas.* Instituto Nacional de Antropología e Historia-Comisión Nacional para el Desarrollo de los Pueblos Indígenas. https://idegeo.centrogeo.org.mx/uploaded/documents/El_patrimonio_biocultural-Eckart_Boege.pdf

Boege, E. (2020, May 26). La minería industrial en territorios bioculturales de los pueblos indígenas. El despojo de los indígenas de sus territorios en el siglo XXI. *La Jornada de Oriente.* https://www.lajornadadeoriente.com.mx/puebla/la-mineria-industrial-territorios-bioculturales/

Bonfil-Batalla, G. (1972). El concepto de indio: una categoría de la situación colonial. *Anales de Antropología, 9,* 105–124. http://www.revistas.unam.mx/index.php/antropologia/article/view/23077/pdf_647

Bueno, C. (2016). *Indigenous. The Wiley Blackwell encyclopedia of race, ethnicity, and nationalism* (1st ed.). Routledge. https://doi.org/10.1002/9781118663202.wberen477

Cárdenas, J. (2013). La minería en México: despojo a la Nación. *Cuestiones Constitucionales. Revista Mexicana de Derecho Constitucional, 28*, 35–74. http://www.scielo.org.mx/pdf/cconst/n28/n28a2.pdf

Cardoza-Zúñiga, R. (2017). Los derechos de los pueblos indígenas y el derecho a un medio ambiente sano: derechos humanos considerados parcialmente en la reforma constitucional en materia energética en México. *Anuario de Derecho Constitucional Latinoamericano, 23*, 243–261. https://www.corteidh.or.cr/tablas/r37890.pdf

Carpio-Domínguez, J. (2021). Crimen Organizado (narcotráfico) y conservación ambiental: el tema pendiente de la seguridad pública en México. *Revista CS, 33*, 237–274. https://doi.org/10.18046/recs.i33.4076

Castro-Salazar, J., & Luyando-Cuevas, J. (2020). Evaluación de la actividad de inspección de recursos forestales en México. *Textual, 75*, 11–35. https://doi.org/10.5154/r.textual.2019.75.01

Centro Mexicano de Derecho Ambiental (CEMDA). (2020). *Informe sobre la situación de las personas defensoras de los derechos humanos ambientales.* https://www.cemda.org.mx/wp-content/uploads/2020/03/informe-personas-defensoras-2019.pdf

Centro Mexicano de Derecho Ambiental (CEMDA). (2021). *Informe sobre la situación de las personas defensoras de los derechos humanos ambientales en México.* https://agua.org.mx/wp-content/uploads/2021/04/informe_cemda_2021_vfinal.pdf

Cherem, A. (2016). *Grupos de autodefensa: sin salida ante la violencia* [tesis de licenciatura, Instituto Tecnológico Autónomo de Mexico]. Repositorio institucional ITAM. http://departamentodecienciapolitica.itam.mx/sites/default/files/u327/tesiscpol_alexischerem.pdf

Coalición para los Derechos Humanos en el Desarrollo (CDHD). (2020). Criminalización, amenazas y ataques contra integrantes de movimientos de pueblos indígenas. Riesgos no calculados. Amenazas y ataques contra las personas defensoras de los derechos humanos y el papel de los financiadores del desarrollo. Caso Nº 2. https://rightsindevelopment.org/wp-content/uploads/2020/06/CASO-2.pdf

Collier, G., & Rus, J. (2002). Una generación de crisis en Los Altos de Chiapas: Los casos de Chamula y Zinacantán, 1974–2000. In S. Mattiace, R. Hernández, & J. Rus (Eds.), *Tierra, libertad y autonomía: Impactos regionales de zapatismo* (pp. 157–199). Centro de Investigaciones y Estudios Superiores en Antropología Social.

Comisión Económica para América Latina y el Caribe (CEPAL). (2014). *Los pueblos indígenas en América Latina. Avances en el último decenio y retos pendientes para la garantía de sus derechos.* Naciones Unidas. CELADE-CEPAL. https://repositorio.cepal.org/bitstream/handle/11362/37222/S1420521_es.pdf

Comisión Nacional de Derechos Humanos (CNDH). (2020). *Derechos de los pueblos indígenas.* https://www.cndh.org.mx/programa/34/pueblos-y-comunidades-indigenas

Comisión Nacional Forestal (CONAFOR). (2013). *Boletín 149.* Gobierno de México. http://www.conafor.gob.mx:8080/documentos/docs/7/48505%20millones%20de%20personas%20ind%C3%ADgenas%20habitan%20en%20bosques%20y%20selvas%20de%20M%C3%A9xico.pdf

Convention on Biological Diversity. (2018). *Conference of the parties 14ª ordinary meeting*. https://www.cbd.int/conferences/2018

Czarny, G. (2012). *Jóvenes indígenas en la UPN Ajusco*. Universidad Pedagógica Nacional.

De Carvalho, S., Goyes, D., & Vegh-Weis, V. (2021). Politics and indigenous victimization: The case of Brazil. *British Journal of Criminology, 61*, 251–271. https://doi.org/10.1093/bjc/azaa060

De Sousa-Santos, B. (2010). *Refundación del Estado en América Latina. Perspectivas desde una epistemología del sur*. Ediciones IVIC. http://www.boaventuradesousasantos.pt/media/Refundacion%20del%20Estado_Lima2010.pdf

Dietz, G. (2017). Interculturalidad: una aproximación antropológica. *Perfiles educativos, 39*(156), 192–207. http://www.scielo.org.mx/scielo.php?script=sci_arttext&pid=S0185-26982017000200192&lng=es&tlng=es

Durin, S. (2007). ¿Una educación indígena intercultural para la ciudad? El Departamento de Educación Indígena en Nuevo León. *Frontera Norte, 19*(38), 63–91. https://www.redalyc.org/articulo.oa?id=136/13603803

Escobar, A. (1998). *La invención del tercer mundo, Construcción y deconstrucción del desarrollo.* Grupo Editorial Norma. https://cronicon.net/paginas/Documentos/No.10.pdf

Esteva, G. (1996). Desarrollo. In S. Wolfgang (Ed.), *Diccionario del Desarrollo: Una Guía del Conocimiento como Poder* (pp. 52–78). PRATEC. https://www.uv.mx/mie/files/2012/10/SESION-6-Sachs-Diccionario-Del-Desarrollo.pdf

Foladori, G. (2001). *Controversias sobre la sustentabilidad. la coevolución sociedad-naturaleza*. Porrúa.

Food and Agriculture Organization (FAO). (2011). *Acumulan pérdidas en bosques densos en México por deforestación.* http://www.fao.org/in-action/agronoticias/detail/es/c/507489/#:~:text=La%20OIMT%20estima%20que%20entre,viven%20en%20condiciones%20de%20pobreza

Gallegos-Martínez, G., & Cruz-Salazar, T. (2021). Identidades juveniles indígenas: influencia de las transiciones a contextos universitarios y urbanos. El caso de Chiapas, México. *CPU-e, Revista de Investigación Educativa, 32*, 33–61. https://doi.org/10.25009/cpue.v0i32.2733

Galván-Meza, C., Pascual-Pérez, S., & Gómez-Anaya, V. (2019). Del despojo, a la identidad diluida, el caso de San Antonio Huitzquilico un análisis sobre el proceso de aculturación a partir del despojo de la lengua Tének. In R. Vázquez; G. Castillo, F. Alejandro & S. Rappo (Coords.), *Desarrollo, disputas socio-territoriales y violencia en México/Disputas socio territoriales en México* (pp. 41–54). Incunabula.

García, M., & Horbath, J. (2019). Exclusión y discriminación de indígenas en Guadalajara, México. *Perfiles Latinoamericanos, 27*(53), 1–24. https://doi.org/10.18504/pl2753-011-2019

Global Witness. (2020). *Defending tomorrow, the climate crisis and threats against land and environmental defenders.* https://www.globalwitness.org/en/campaigns/environmental-activists/defending-tomorrow/

Gobierno de México. (2013). *Explicación ampliada de la Reforma Energética.* https://www.gob.mx/cms/uploads/attachment/file/10233/Explicacion_ampliada_de_la_Reforma_Energetica1.pdf

Gobierno de México. (2014). *Programa Nacional Forestal 2014–2018.* Gobierno de México. http://www.conafor.gob.mx:8080/documentos/docs/4/5382Programa%20Nacional%20Forestal%202014-2018.pdf

Gobierno de México. (2020). *Pueblos indígenas. Sistema de Información Cultural.* Gobierno de México. https://sic.cultura.gob.mx/index.php?table=grupo_etnico

González-Montes, S. (2009). Violencia contra las mujeres, derechos y ciudadanía en contextos rurales e indígenas de México. *Convergencia, 16*(50), 265–1185. http://www.scielo.org.mx/pdf/conver/v16n50/v16n50a7.pdf

Goyes, D. (2017). Corporate lobbying and criminalization. *Crime, Law & Social Change, 69*, 401–419. https://doi.org/10.1007/s10611-017-9745-8

Goyes, D. (2019). *Southern green criminology: A science to end ecological discrimination.* Emerald Publishing.

Goyes, D. (2021). Environmental crime in Latin America and Southern Green Criminology. *Oxford Research Encyclopedia of Criminology. Criminology and Criminal Justice.* https://doi.org/10.1093/acrefore/9780190264079.013.588

Goyes, D., & South, N. (2021). Indigenous worlds and criminological exclusion: A call to reorientate the criminological compass. *International Journal for Crime, Justice and Social Democracy, 10*(3), 115–128. https://doi.org/10.5204/ijcjsd.1983

Goyes, D., South, N., Abaibira, M., Baicué, P., Cuchimba, A., & Ramos, D. (2021). Genocide and ecocide in four Colombian Indigenous Communities: The Erosion of a way of life and memory. *The British Journal of Criminology, 61*(4), 965–984. https://doi.org/10.1093/bjc/azaa109

Haesbaert, R. (2011). *El mito de la desterritorialización. Del fin de los territorios a la multiterritorialidad*. Siglo XXI Editores.

Hernández-Moreno, J., & Nuñez-Vera, M. (2014). Conservación de recursos forestales y género, en el marco del desarrollo sustentable. El caso de la comunidad indígena barrio de San Miguel, Michoacán, México. *Revista*

Cubana de Ciencias Forestales, 2(2), 1–20. https://dialnet.unirioja.es/servlet/articulo?codigo=5223130

Instituto Nacional de Estadística y Geografía (INEGI). (2020a). *Estadísticas a propósito del Día Internacional de los pueblos indígenas (9 de agosto)*. Comunicado de Prensa Núm. 392/2020a. Gobierno de México. https://www.inegi.org.mx/contenidos/saladeprensa/aproposito/2020a/indigenas2020a.pdf

Instituto Nacional de Estadística y Geografía (INEGI). (2020b). *Hablantes de lengua indígena en México*. Gobierno de México. http://cuentame.inegi.org.mx/poblacion/lindigena.aspx#:~:text=En%20M%C3%A9xico%207%2C364%2C645%20personas%20de,ind%C3%ADgena%2C%2012%20no%20hablan%20espa%C3%B1ol.

Instituto Nacional para los Pueblos Indígenas (INPI). (2020). *Atlas de los pueblos indígenas de México*. Gobierno de México. http://atlas.inpi.gob.mx/

López Bárcenas, F. (2002). Territorios, tierras y recursos naturales de los pueblos indígenas en México. In J. González (Coord.), *Constitución y derechos indígenas* (pp. 121–143). UNAM.

Madrid, L., Núñez, J., Quiroz, G., & Rodríguez, Y. (2009). La propiedad social forestal en México. *Investigación Ambiental, 1*(2), 179–196. https://biblioteca.semarnat.gob.mx/janium/Documentos/Ciga/agenda/DOFsr/INVEAMB000016.pdf

Martínez-Coria, R., & Haro-Encinas, J. (2015). Derechos territoriales y pueblos indígenas en México: una lucha por la soberanía y la nación. *Revista Pueblos y Fronteras, 10*(19), 228–256. http://www.scielo.org.mx/pdf/rpfd/v10n19/1870-4115-rpfd-10-19-228.pdf

Martínez-Corona, B. (2003). Género, sustentabilidad y empoderamiento en proyectos ecoturísticos de mujeres indígenas. *La Ventana Revista de Estudios de Género, 2*(17), 188–217. https://doi.org/10.32870/lv.v2i17.648

Melucci, A. (1999). *Acción colectiva, vida cotidiana y democracia*. El Colegio de México.

Moreno, R. (2010). Indígenas en la ciudad y en la lejanía. la representación diferenciada de los indígenas en el discurso periodístico (El Norte: 1988–2006). En S. Durin (Ed.), *Etnicidades urbanas en las américas. procesos de inserción, discriminación y políticas multiculturalistas* (pp. 253–268). Publicaciones de la Casa Chata.

Ocampo-Rodríguez, D. (2015). Porous borders: Illegal mining and drug trafficking as latent threats to security. *Management and Strategy, 10*(1), 141–150. https://dialnet.unirioja.es/servlet/articulo?codigo=5682851

Ortiz-Hernández, L., Ayala-Guzmán, C., & Pérez-Salgado, D. (2018). Posición socioeconómica, discriminación y color de piel en México. *Perfiles*

Latinoamericanos, 26(51), 215–239. https://doi.org/10.18504/pl2651-009-2018

Paz-Salinas, M. (2011). Luchas en defensa del territorio. Reflexiones desde los conflictos socio ambientales en México. *Acta Sociológica, 73*, 197–219. https://doi.org/10.1016/j.acso.2017.08.007

Pérez-Jiménez, S. (2014). *Territorialidades contenciosas en México: el caso de la minería* [tesis de maestría, Universidad Nacional Autónoma de México]. Repositorio institucional UNAM. https://comisiones.ipgh.org/CARTOGRAFIA/Premio/Tesis_2015/Tesis_Sol_Perez.pdf

Pérez-Ruiz, M. (2019). Jóvenes indígenas en América Latina: Reflexiones para su investigación desde la antropología. *Anuario Antropológico, 2*, 21–50. https://doi.org/10.4000/aa.4003

Pérez, R. (2015). Jóvenes indígenas como actores sociales con identidades múltiples. En M. Ruiz, V. Ruiz y S. Velasco (Eds.), *Interculturalidad(es). Jóvenes indígenas: educación y migración* (pp. 337–360). Universidad Pedagógica Nacional.

Reyes-Aguilar, A. (2017). *Mujeres, conservación ambiental y turismo rural en el ejido Adolfo López Mateos, Los Tuxtlas, Veracruz* [Thesis. Universidad del Estado de México]. Institutional Repository. http://hdl.handle.net/20.500.11799/68822

Rodríguez-Wallenius, C. (2017). Disputas territoriales en torno a las plantaciones forestales y de agrocombustibles en el sureste de México. *El Cotidiano, 201*, 59–66. https://www.redalyc.org/pdf/325/32549629007.pdf

Rodríguez, N. (2002). *Pueblos indios, globalización y desarrollo (ensayo para entender).* Instituto Nacional Indigenista-PNUD. https://www.nacionmulticultural.unam.mx/portal/pdf/cultura_politica/nemesio_rodriguez_20081105.pdf

Romero-Zepeda, H., & Ortega-Marín, B. (2017). Vulnerabilidad impuesta y violación sistemática a los derechos de las mujeres indígenas en México. *Revista Médica Electrónica, 39*(4), 933–946. http://scielo.sld.cu/pdf/rme/v39n4/rme080417.pdf

Rousseau, I. (2019). La reforma energética (2013–2014) a la luz de la nueva legislación sobre los impactos sociales de los proyectos. *Foro Internacional, 60*(240), 853–887. https://doi.org/10.24201/fi.v60i2.2740

Ruiz-Rincón, V. (2015). Los pros y los contras de la reforma energética de acuerdo con el paradigma del desarrollo sustentable. *Revista Digital Universitaria, 16*(1), 1–16. http://www.revista.unam.mx/vol.16/num1/art08/art08.pdf

Sainz, J. (2018). Narco Deforestación en México. Un vínculo ambivalente. In L. Atuesta & A. Madrazo (Eds.), *Las Violencias, en busca de la política pública detrás de la guerra contra las drogas* (pp. 51–74). Coyuntura y Ensayo/ CIDE.

Soberanes, R. (2020, May 25). *Eugui Roy Martínez: la increíble vida del biólogo asesinado en Loxicha.* https://es.mongabay.com/2020/05/eugui-biologo-asesinado-en-mexico/

Solís, P. (2017). *Discriminación estructural y desigualdad social. Con casos ilustrativos para jóvenes indígenas, mujeres y personas con discapacidad.* Consejo Nacional para Prevenir la Discriminación. http://repositorio.dpe.gob.ec:8080/bitstream/39000/2084/1/CONAPRED-066.pdf

Sozzo, M. (2003). ¿Contando el delito? Análisis crítico comparativo de las Encuestas de Victimización en Argentina. *Cartapacio de Derecho: Revista virtual de la Facultad de Derecho, 5*, 1–143. https://dialnet.unirioja.es/servlet/articulo?codigo=5028431

Toledo, V., Garrido, D., & Barrera-Bassols, N. (2014). Conflictos socioambientales, resistencias ciudadanas y violencia neoliberal en México. *Ecología política, 45*(41), 115–124. https://www.ecologiapolitica.info/?p=1266

Toledo, V., Garrido, D., & Barrera-Bassols, N. (2015). The struggle for life, socioenvironmental conflicts in México. *Latin American Perspectives, 42*(5), 133–147. https://doi.org/10.1177/0094582X15588104

Tetreault, D., Ochoa-García, H., & Hernández-González, E. (2012). *Conflictos socioambientales y alternativas de la sociedad civil.* Instituto Tecnológico y de Estudios Superiores de Occidente. https://rei.iteso.mx/bitstream/handle/11117/425/Conflictos%5B1%5D.pdf?sequence=2

Valladares-De la Cruz, S. (2017). El despojo de los territorios indígenas y las resistencias al extractivismo minero en México. *e-cadernos CES, 28,* 21–45. https://doi.org/10.4000/eces.2291

Vázquez-García, V. (2003). La gestión ambiental con perspectiva de género. El manejo integrado de ecosistemas y la participación comunitaria. *Gestión y Política Pública, 12*(2): 291–322. https://www.redalyc.org/pdf/133/13312204.pdf

Vázquez-Parra, J., & Campos-Rivas, C. (2016). Discriminación laboral indígena: una aproximación desde el imaginario colonial y la Teoría Elsteriana. *Saber, 28*(4), 828–837. http://ve.scielo.org/pdf/saber/v28n4/art18.pdf

Velázquez, A. (2017). *Desplazamiento interno por violencia en México. Causas, consecuencias y responsabilidades.* Comisión Nacional de los Derechos Humanos. http://www.corteidh.or.cr/tablas/r37820.pdf

Villavicencio-Enríquez, M. (2011). Conservation of the montes Azules biosphere reserve, Chiapas: A women's issue. *Development, 54*(4), 473–479. https://doi.org/10.1057/dev.2011.91

Wallerstein, I. (2004). *World-systems analysis: An introduction.* Duke University Press.https://doi.org/10.2307/j.ctv11smzx1

World Bank. (2020). *Indigenous peoples.* https://www.bancomundial.org/es/topic/indigenouspeoples

5

Appropriating the Commons: Tea Estates and Conflict Over Water in Southern Malawi

Dave Mankhokwe Namusanya

Introduction

In Malawi's Thyolo and Mulanje districts, tea estates are a major feature. The districts have earned a reputation as beautiful tourism spots owing to the vast green of the tea plantations. It is a greenery that usually survives the harshest of summers when all lands around are brown. This is usually due to sprinklers that keep irrigating the plantations. Sprinklers, however, clash with the reality of thirsty communities in the margins of the tea fields. Yet this unyielding scenery of vast tea fields continues for kilometres, almost reaching the Mulanje District headquarters, and pushes all the way into the area that borders Mozambique. As they take up the land, the tea estates push into further distance local communities, cutting them off from important amenities including roads and healthcare services.

D. M. Namusanya (✉)
Abertay University, Dundee, UK
e-mail: d.namusanya1900@abertay.ac.uk

D. R. Goyes (ed.), *Green Crime in the Global South*, Palgrave Studies in Green Criminology, https://doi.org/10.1007/978-3-031-27754-2_5

A colonial legacy, the tea estates are mostly owned by British corporations—an issue that in recent years has caused consternation among locals. While tourists celebrate the tea estates as a natural relic that adds to the beauty of the districts, locals look at them as encroachments. They accuse them of taking up land for farming. Thus, in more recent years, the estates have come under the scrutiny of local people. Indigenous activists have been making demands that some, or all, of the land be returned to them (Kishindo & Mvula, 2017; Mangulama & Wu, 2022).

A growing interest in land both for farming and settlement has placed the issues of the tea estates and access to land in Malawi in the spotlight (Namangale, 2021). There have not been any violent land seizures yet as the People Land Organisation (PLO), a movement that took it upon itself to lead the land reclamations, puts away the possibility of violent confrontation. This, however, is not to say that there might not be violent confrontations. The PLO has been pushing for an independent republic. Its leader, Vincent Wandale, declared himself supreme leader of that republic in 2015. His agitations, nevertheless, are non-violent. His subsequent diagnosis with psychosis also negatively affected the mobilisation he leads. On the ground, however, my field observations in Mulanje District between February and November 2021, revealed that the need for land redistribution is strong.

The Malawi Government has often been called upon to act on the land issues. They have since attempted to address the issues with the support of multinational institutions such as the World Bank. Thus, in 2004, the *Kuzigulila malo* (vernacular for self-acquisition of land) project was launched. This project aimed to ease land pressures on Mulanje and Thyolo districts by relocating the people from the districts into other districts within the country (Mangulama & Wu, 2022). Its success, however, has been blighted by other realities. One of the issues that the project has had to deal with has been that geographical spaces in Malawi are, for the most, the ancestral lands of particular Indigenous communities. Districts have strong ties with specific aboriginal peoples. For instance, the Lhomwe-Mang'anja people are regarded as the ancestral inhabitants of the Mulanje and Thyolo districts. The Yao, on the other hand, are regarded as traditional inhabitants of Mangochi and Machinga where people were relocated into from Mulanje. Relocations created tension based on feelings of attachment to their territories. Conflicts over land therefore keep smouldering.

As much of the focus on conflicts in Mulanje and Thyolo districts is currently over land, future conflicts are to increasingly be on the complex relationships of land and water. Water will be a resource where struggles and contestations will manifest. This chapter is based on ethnographic fieldwork conducted in Mulanje District for my doctoral studies. It shows how colonialism altered human–water relationships in this district. The chapter further highlights how colonialism in its different manifestations—even under the guise of *development* (Escobar, 1995)—still impacts on these relationships. Colonialism in this chapter is regarded as a form of environmental harm and a criminogenic process (see, for example, Goyes, 2021).

The underlying premise of this chapter is that water is a common whose use and relevance in societies rests on broader historical connections that humans have towards the resource. Such connections override recent consumerist and neo-colonial perspectives—both products of the colonial reordering of the world. Water in this context is taken as a *complex social phenomenon*, as suggested by Orlove and Caton (2010) as well as other scholars (see, for example, Attala, 2016; Ballestero, 2019; Strang, 2004). In this way of thinking, water is not just a physically flowing resource. It is rather a resource with social meaning. It is at the centre of human relationships, even the human and the supernatural worlds (Orlove & Caton, 2010; Strang, 2004).

The chapter starts by highlighting the framework of Southern green criminology. This is mostly concerned with the contextual particularities of the location where violence takes place. I then offer the evidence of the green crimes committed in each of the distinct periods from colonisation up to today regarding tea crops. The chapter proceeds by highlighting the context of the area under focus, Mulanje. In this section, I highlight the nature of green crimes committed in the area. The chapter also offers explorations of possible scenarios for the future, particularly in the contexts of climate change and the pervasive nature of colonialism that has redefined human–environmental relationships in the global South.

Operationalising Southern Green Criminology

Green criminology, which is the bedrock of the criminology focused on in this chapter, has garnered noteworthy attention in recent years. This has mostly been inspired by the change in direction of criminology, which used to focus on Western realities and ways of knowing, but has recently started paying increased attention to the impacts of colonialism (Goyes, South, Abaibira et al., 2021). From the turn of the millennium when green criminology was mostly tied to the 'prosecutable' harms on the environment (see Halsey, 2004), green criminology now gains considerable research attention in which the acts of state and non-state actors sit at the centre of examination despite many of them being legal (Lynch, 2020; White, 2021).

Green criminology thus studies phenomena that range from understanding the impacts that humans and corporations (including the state) have over the environment to understanding the harmful impacts on people and other living organisms. These harms come through the degradation of the environment. This chapter, by tracing the harmful impacts on the environment of violent colonial reframing of the social order, focuses on the impacts that green crimes have had on the people that live with and rely on nature for their survival (as opposed to people who have distanced themselves from nature, only receiving its goods). This chapter is inspired by recent academic interests in Southern green criminology. Goyes (2021) indicates that Southern green criminology arises out of the need to give enough consideration to the role of colonisation in environmental harm. Thus, in furthering the argument, I focus on the geopolitical divisions of North/South in the creation and perpetuation of harm both to the environment and humans (mostly the Indigenous and the poor) reliant on it.

Southern green criminology has inspired this work by highlighting the importance of the spatiotemporal context for understanding environmental harm, including the trajectories and legacies of colonialism and neo-colonialism. Southern locations have, for the most, three main epochs: the pre-colonial period, the colonial period, and the neo-colonial period (following the conceptualisation offered by Castro Herrera, cited in Goyes [2021]). In this chapter, I extend the discussion with a fourth, futuristic consideration: roughly called the Anthropocene post-modern epoch. This is influenced by the reading of current trajectories to forecast a future environmental epoch which is not only shaped by colonial-capitalist interests; it is also moulded by realities of climate change (this is similar to thoughts and arguments advanced by Brisman et al. [2018] and White [2021]).

Under this framework of Southern green criminology, this chapter focuses on the nature of human–water relationships that existed prior to colonisation (the pre-colonial); how the British control of Malawi (then Nyasaland) changed the nature of such relationships (the colonial); and, importantly, the era of neo-colonialism—in which Western capitalism orders and shapes the relationships that exist between humans and water (the post-colonial). I further highlight potential altered changes in a climate-impacted future (the Anthropocene post-modern). The discussion in the chapter will also focus on the role that capitalism has in neo-colonial social orders. The focus of the discussion will be on how capital is more valued than people's traditional relationship to water and the environment at large (Mulwafu, 2011; Shumba, 2011).

As the chapter shows, there is an overlap of epochs. The pre-colonial epoch, with its emphasis on traditional human–nature relationships, still maintained a hold on social life in the colonial epoch. As McCracken (2012) points out, colonisation did not completely uproot the social life in Malawi. There was still resistance, a pushback, against the violence of colonialism. This pushback permeates within the post-colonial such that ideals of the precolonial epoch maintain a footing. Even in the post-modern epoch, this overlap can be expected. However, precolonial relationships of humans to the environment are increasingly under threat (Goyes, South, Abaibira, et al., 2021). This threat is imposed either through a legal violence of appropriation of what would be regarded as the commons, or a structural violence imposed by the economic systems of capitalism and development (Escobar, 1995; Nixon, 2011).

The understandings of Southern green criminology that underpin this chapter therefore situate the concept of decolonisation as a particular frame of reference. Decolonisation is understood as the undoing of the centrality of Western knowledge. The goal is to create a new 'epistemology' that considers Indigenous knowledges and realities as valid (Goyes, 2021; Goyes, South, Sollund, et al., 2021; Ndlovu-Gatsheni, 2015). It is a rejection of the genocidal and ecocidal logics brought about by colonialism (Goyes, South, Abaibira, et al., 2021).

The Nature of Colonialism in Malawi

It is important to describe the nature of colonialism that Malawi experienced ahead of discussing its impacts on modern life. Colonialism, with all its attendant violence, was not imposed in the same manner throughout the global South. For instance, the colonisation of Malawi was mostly driven by the colonial interests of agriculture and labour (Kalinga, 1996; Potts, 2012; Tchuwa, 2019). It became a British protectorate in 1907 having had initial contact with colonialists through David Livingstone (whose impact on Malawi looms large even today) in 1859. Then, Malawi was called Nyasaland. Unlike a colony, the status of Malawi meant that it was regarded as a 'colonial backwater' (Tchuwa, 2019)—a place the British Empire had little thought for and were only interested in its resources of land, water and labour. As McCracken (2012) records, the colonial administrative policies that were imposed in Nyasaland were certainly different from those imposed on other colonies in Southern Africa, most notably Southern and Northern Rhodesia (now Zimbabwe and Zambia) for which Malawi was to later be conjoined into a federation.

While colonialism in other parts was driven by the need for minerals, in Southern Africa it was also driven by the need for agricultural land. Potts (2012) has highlighted that the prospects of minerals in some parts of Southern Africa were little such that colonialists turned to farming. This interest in farming eventually meant that they looked for fertile land. McCracken (2012) points out that this search often led colonists to areas that received good amounts of rainfall—they sought land with access to suitable amounts of rainfall for agriculture, and reliable water sources. With this background begins the troubled history of water in

Southern Africa. It is a history that carries on today and that will probably exacerbate conflict in the near future as climate change worsens the socio-environmental conditions.

Precolonial Human–Water Relationships

Historians highlight the role that water played in structuring pre-colonial Malawian society (McCracken, 2012; Tchuwa, 2019). Water was especially relevant in the spirituality of Indigenous peoples. John McCracken points out that it was at the centre of religious practices for Indigenous people (McCracken, 2008). Tales of the rainmaker, Makewana, emanating from the recesses of history and within the Malawian folklore narratives, point to this relationship with water. Tchuwa (2019) details the nature of human–water relationships in precolonial Southern Malawi stressing that they were mostly exercised under the rubric of religion. *Rainmaking*, or activities in which royal family members would perform traditional dances or offer sacrifices to call for rains, were a significant feature in societies. Indeed, the success of colonists among the Ngoni people of Northern Malawi has also been alluded to their ability to have 'made rains fall after a drought' (McCracken, 2008, p. 59). This, of course, would be attributed to their ability to read the weather.

Similarly, Mulwafu (2011) indicates that the relationship between humans and the environment in precolonial times was significantly shaped by humans due regard of the environment. In this regard, cultural, social, and *religious* rites played a key role in ordering such relationships. Observing a human connection with the environment, Hyam (2010) describes the animistic religions common in Africa before colonialism. People worshipped spirits believed to inhabit the environment. However, as Mulwafu (2011) and Shumba (2011) indicate, these were not worship styles centred on worshipping the environment; rather, they were rooted in the belief that the environment is as valuable as humans. In my ethnography, community members reported the same stressing that their ancestors regarded water as the centre of their social life (see, also, McCracken, 2012). Pre-colonial societies had an affinity to the environment in which they regarded it a source of human life.

Water, therefore, was equally deserving of respect as the rest of the earth (Shumba, 2011). This was different to the way that colonists looked at the environment. Sharp (2008) has argued that colonists often looked at such human–environment relationships with disdain. The relationship between the aboriginal inhabitants of Malawi and water, for instance, drove colonialists to assert that Indigenous peoples were inferior because they considered themselves to be part of the environment. Thus, it was held that in violently dominating such people they were dominating the environment. Indigenous people were regarded as the wild. The colonial crusade was premised on Christianity (see McCracken, 2012), and the Bible verse[1] stating that god gave humans dominion over all the earth inspired the colonisers (these sentiments have also been highlighted by Shumba [2011], in his work on the African philosophy of Ubuntu and the environment).

Tchuwa (2019) argues that such condescending attitudes of the British led almost to the erasure of local relationships to water. In their pursuit of the 'sanitation of the uncivilised blacks'—who I dare add were regarded as wild creatures, the same as all others in nature—the colonisers appropriated *the commons* to advance their commercial purposes. The commercialisation of water has indeed been highlighted by other scholars as being a Western idea. Strang (2004), for instance, has pointed out that the regard of a duality in humanity's relation to water (i.e. as two separated elements rather than as a unity) has mostly been prevalent in Western societies. Precolonial relationships to water were not based on difference or dominion over the resource, but on living in harmony with it (Mulwafu, 2011).

[1] In the Bible, Genesis 1:26–28, states: 'Then God said, "Let us make mankind in our image, in our likeness, so that they may rule over the fish in the sea and the birds in the sky, over the livestock and all the wild animals and over all the creatures that move along the ground." So God created mankind in his own image, in the image of God he created them; male and female he created them. God blessed them and said to them, "Be fruitful and increase in number; fill the earth and subdue it. Rule over the fish in the sea and the birds in the sky and over every living creature that moves on the ground"'.

Relationships to Water in the Colony

Water facilitated colonialism (Hauser, 2021; McCracken, 2012; Sharma, 2022). Water's 'guilt' in the processes of colonialism is not only attuned to its transportation qualities of carrying the vessels of the colonisers. It also—and most importantly for this chapter—created a clear division between the colonisers and the colonised depending on the ways of relating and the access to it. Water created spaces labelled as urban, civilised, and white (Riley, 2012), which were deemed as 'outside' for the uncivilised and Indigenous (Tchuwa, 2019). Thus, in Malawi, areas with good water sources and clear water (McCracken, 2012) were appropriated by colonial settlers. Indigenous people were pushed into areas where water access was difficult (Tchuwa, 2019).

Cultural colonialism means imposing a way of being over the colonised (see Goyes, 2021). Thus, the very essence of colonialism and its claims to civilisation would hardly have existed without the mutual conquering and 'civilising' of water or—to quote Tchuwa (2019)—without the 'capture and subjugation of water to ... the colonial legacies of transforming society–nature–space relations' (p. 71). When Malawi became a British colony, native relationships with water changed. Water became exclusively a biophysical resource, and its main use outside the biological function of sanitation became irrigation (Mathur & Mulwafu, 2018). The relationships of humans to water, which were spiritual, became mediated through capitalistic interests (Tchuwa, 2019). Water was commodified (Castree, 2003).

Tchuwa (2019) indicates that colonisers regarded water as purely a biological and agricultural entity. Thus, a criminogenic relationship between water and humans was created. In this newly defined relationship, water became a commodity governed and distributed along the markers of race and power (Mathur & Mulwafu, 2018). This was especially true for Blantyre City that was the hub of colonial activities in Southern Malawi.

In the violence of colonialism, the criminogenic relationship between humans and water was further witnessed in the 'Christianisation' of Indigenous people. When colonisers 'civilised' (and christened) natives, ceremonies of rainmaking were erased. In my ethnography, community members in Mulanje District recollected stories of their ancestors being stopped from using water resources for religious functions other than baptism. Indeed, significant religions in Africa disappeared upon the spreading of Christianity (Hyam, 2010).

In Mulanje District, the interference that colonialism had on water was not just on religion. It was also on agriculture. This was due to the fact that the area was chosen for the cultivation of tea and coffee. Morris (2016) highlights that the weather of the area was good, both for settlement and agriculture. McCracken (2012) records that fertile soils and the availability of rivers with clear water made Mulanje attractive to colonisers. It was on these lands, closely twinned to rivers, that much of the colonial activities in the district happened (Kalinga, 1996). Even if rainmaking ceremonies survived at the Dziwe la nkhalamba pool ('pool for the elderly'), colonialism significantly altered relationships to water through the introduction of new crops and agricultural practices.

Current Human Relationships to Water

Mulanje District is one district where water is regarded as abundant, with at least nine major rivers flowing across the Mulanje Mountain massif (Nangoma & Nangoma, 2013). In a general overview of the water situation in Malawi, it is the case that water is regarded as *a given* (Barnes, 2014). That is, its availability is assumed to be guaranteed. Malawi relies on the freshwater of Lake Malawi. The existence of big rivers flowing copiously throughout the year has buttressed the impression that Malawi has enough, even too much, water. Brown (2017) indicates that the physical presence of water sources in an area often conjures an image of water abundance. This is certainly true in Malawi, particularly in Mulanje.

However, this image is increasingly being shattered. The rain season is getting increasingly erratic (Gallo, 2018) and water levels in rivers are dropping (Mtilatila et al., 2020). Water availability is under threat. Climate change, an increasing population, and urbanisation all exert pressure on water sources (Tchuwa, 2017). Furthermore, capitalism and its innate exploitation of resources has kept impacting on water. Water in the Anthropocene post-modern epoch is no longer a common— it has been commodified with a significant market value (Tchuwa, 2019). Its distribution, and use, is a perpetuation of the colonial structures of power. The only difference, as Tchuwa (2019) argues, is that race has been replaced with class. The well-off who took over areas initially occupied by white settlers have access to water.

Despite these transformations brought about by colonialism, narratives of human–water relationships in Mulanje are still embedded in spiritual and cultural understandings. Locals still indicate the historical significance of some ponds and water spots on Mulanje Mountain (Malijani, 2019). Nevertheless, the colonial legacy of seeking to control water and manage it for commercial purposes affects the way water is viewed in Mulanje District. In recent years, the water of Mulanje has started attracting the attention of the Malawian state which has been appropriating it with little regard for the Indigenous views about the water in the area (see Namusanya et al. [2022] on a state water project in Mulanje District). The state has become an accomplice in this blue crime (Lynch, 2020).

At the moment of writing, on the Ruo and Ndiza rivers on Mulanje Mountain, Mulanje Hydro Limited produces electricity for the Malawian energy company, EGENCO. The water has now become a national commodity on which communities have little say (Hoag, 2019). As Castree (2003) indicates, commodification perpetuates exclusion and harm on the people and/or the environment. Thus, any intention of using and managing water in ways that run contrary to the expectations of the state often leads to clashes between the state and local communities. Communities have complained that when making national water management policies, they, their needs, and their interests have been overlooked. They have also accused the state of using physical violence to repress them (Pensulo, 2018). The state shields commercial interests and perpetuates harms to the environment and the people living off it (Goyes, 2021).

A more troubling outlook is the perpetuation of the colonial legacy of tea farming (Mangulama & Wu, 2022). The estates are a symbol of Malawi's status as a 'colonial other' as the Indigenous inhabitants increasingly become the *subjects* of such a system (the categorisations are inspired by the work of Mathur and Mulwafu [2018], who referred to the work of Mamdan). Water sits at the centre of these competing interests with tea estates needing the water for irrigating the tea fields while the communities around it need it for daily sustenance. The communities also need it for other uses such as tourism (that benefits the local people) as well as the historical and spiritual connotations that it has usually had among them (McCracken, 2012).

Communities in Mulanje complain that the tea estates' appetite for water has been increasing over the years. According to them, the tea estates introduced a new brand of tea crop which needs irrigation almost all year round. The demand for irrigation has further been exacerbated by the unpredictability of rainfall in Malawi (Wilson, 2018). Research on tea farming in Mulanje has indeed indicated that, as climate change impacts worsen, tea will demand more irrigation (Bouwer et al., 2021). This will increase pressure on water sources. As the tea estates have the resources and the machinery to extract more water, they control and govern the relationship with water bodies as well as the communities around. Water technologies do not always foster equality (Strang, 2004). In this context and contestation, water is no longer the common that people knew but has instead become a commodity for which individual actors compete.

Capitalistic ways of appropriating water compete with traditional populations not only in accessing water physically, but also in the ways it should be conceptualised. People in rural Malawi still want to regard water as a common (Cleaver et al., 2021). In the rural communities of Mulanje, people in areas under different traditional leadership live off the same streams. Water management here, often outside of the purview of non-governmental organisations (NGOs) promoting their own approach to resource management, is a communal practice. This is mostly rooted in the historical and cultural realities of sharing and regards to common interest as well as spiritual values (Malijani, 2019). These, nevertheless, remain under the constant threat of colonial notions of water. Increasingly, there is pressure to commodify the water in these areas. Tea estates also play a role in transforming understandings of water as a mere physical resource.

Future Considerations of Human–Water Relationships in Mulanje

The relationships that humans have with water in Malawi, and specifically Mulanje District, have been changing over the years. These changes have been marked by the prevalent economic, political, and social conditions. Since colonisation, the relationship has been marked by a shift towards commodification (Mathur & Mulwafu, 2018; Tchuwa, 2019). Capital has come to define the relationship between humans and water. Water no longer serves local people but has become a commodity controlled by state powers advocating for *development* and by economic actors defending their profiting interests—both execute acts of colonialism. Climate change does not weaken these forces. It emboldens them by giving them an excuse to push forward their policies under the excuses of developmental work and national state interests (Escobar, 1995; Sultana, 2022). Local communities are further deprived of the power to control and manage their resources.

Research has showed that Mulanje is suffering the impacts of climate change more than many regions in the world (Nangoma & Nangoma, 2013; Wilson, 2018). Community members cited unreliable rainfall patterns as well as cyclones as some of the impacts in the district (see, also, Davies, 2022). With rivers and human life being reliant on rains, the changes in rainfall patterns will mean that more pressure will be placed on the available water sources. This will keep having an impact on the way that humans relate with water in the district. Interest will be in securing access to the water resources. At the same time, these will be impacted by the incessant desire of water by the tea estates (Bouwer et al., 2021). There is evidence that when commercial interests conflict with Indigenous peoples' interests, the state caters to the commercial interests. In Mulanje, this has already been witnessed over land (Mangulama & Wu, 2022) and water as well (Namusanya et al., 2022). In this context, green crimes can only be expected to worsen. Climate change will certainly distort human–water relationships. It will also further worsen the position of the Indigenous people in the area. Most of them have little bargaining power. The colonial imposition of capitalism has forced them to live at the mercy of the tea estates. They have to get jobs from the same structure that threatens their existence.

However, it should be pointed out that Indigenous people in Mulanje are not passive subjects of capitalist violence. Often, they have mobilised to challenge it. The outcome has often been bloody with several deaths being reported among locals in their conflicts with tea estates (see, for example, Jimani, 2021). In these confrontations, the state rarely addresses the violence of estates. Remarks by community members that the estates are 'gods in themselves' are often reflected in altercations with local people. The colonial structures, embedded within the Malawi state, protect the capital interests in the estates. Thus, crimes to the environment often cascade into crimes of physical aggression to locals. In other words, slow and structural violence leads to direct and physical violence (Nixon, 2011).

As discussed in the context of Southern green criminology, a new reality hovers on the horizon. In Malawi, not only do the current configurations follow the colonial order, but also the future shape of society will be influenced by neo-colonialism (see McCracken, 2012; Riley, 2012; Tchuwa, 2019). As climate change impacts worsen, and the needs of consumer capitalism keep increasing, the harm on communities living off the environment can be expected to worsen. In this context, there is a sore need for a deliberate undoing of the structures of neo-colonialism for the environmental harms to stop. This, also, would need the state to acknowledge the harm its own policies, added to the agendas of multinational corporations, are having on the environment and its peoples.

Conclusion

The notions of development imposed by the global North on the global South often lead to harm for local communities (Escobar, 1995). In Malawi, water has become a resource through which harm has been done, ranging from the erasure of Indigenous affiliations to water in colonial Malawi to the post-colonial state that has made water a national resource, thereby disenfranchising local people (Hoag, 2019). All this to support the interests of colonialism remodelled as capitalism. This has dispossessed local communities. As the pressure of capitalism increases and the impacts of climate change worsen, more green crimes will transpire in the global South.

Only by decentring Western understandings of crime, common in Malawi, can social justice be realised. As it stands, without conceptualising green crime within the geopolitical space in which it is practised, little will be achieved. The colonial history of violence needs to be centred in the conceptualisations of green criminology. Environmental violence in the global South—encompassing ecocide and genocide of Indigenous people and their ways of living—should be understood as part of the consequences of colonisation (Goyes, South, Abaibira, et al., 2021). Beyond this conceptualisation, there should be deliberate action to prevent further violence against local people. The adaptation to climate change in the global South should not further victimise communities. Interests of capital, for instance on water, should not trump Indigenous interests.

References

Attala, L. (2016). Bodies of water: Exploring water flows in rural Kenya. In *Exploring the materiality of food 'stuffs'* (pp. 95–116). Routledge.

Ballestero, A. (2019). The anthropology of water. *Annual Review of Anthropology, 48*(1), 405–421. https://doi.org/10.1146/annurev-anthro-102218-011428

Barnes, J. (2014). *Cultivating the Nile: The everyday politics of water in Egypt*. Duke University Press.

Bouwer, R., Arajuo, J., Mittal, N., & Watkiss, P. (2021, October 12). *Managing long term risks in the tea sector in Malawi, Kenya and Rwanda*. Africa Portal; Future Climate for Africa (FCFA). https://www.africaportal.org/publications/managing-long-term-risks-tea-sector-malawi-kenya-and-rwanda/

Brisman, A., South, N., & Walters, R. (2018). Southernizing green criminology: Human dislocation, environmental injustice and climate apartheid. *Justice, Power and Resistance, 2*(1), 1–21.

Castree, N. (2003). Commodifying what nature? *Progress in Human Geography, 27*(3), 273–297. https://doi.org/10.1191/0309132503ph428oa

Cleaver, F., Whaley, L., & Mwathunga, E. (2021). Worldviews and the everyday politics of community water management. *Water Alternatives, 14*(3), 645–663.

Davies, R. (2022, March 15). Malawi—4 dead after Cyclone Gombe causes floods in South. *FloodList*. https://floodlist.com/africa/malawi-cyclone-gombe-floods-march-2022

Escobar, A. (1995). *Encountering development: The making and unmaking of the third world*. Princeton University Press.

Gallo, I. (2018). *Between Kasache and Geneva: The multi-sited voyage of climate-resilient development in Malawi* (PhD, Lancaster University; Application/pdf). http://www.research.lancs.ac.uk/portal/en/publications/between-kasache-and-geneva(68a720b1-03d4-4b4f-ad87-296f6bd4a407).html

Goyes, D. R. (2021, March 25). *Environmental crime in Latin America and southern green criminology*. Oxford Research Encyclopedia of Criminology and Criminal Justice. https://doi.org/10.1093/acrefore/9780190264079.013.588

Goyes, D. R., South, N., Abaibira, M. A., Baicué, P., Cuchimba, A., & Ñeñetofe, D. T. R. (2021). Genocide and ecocide in four Colombian indigenous communities: The erosion of a way of life and memory. *British Journal of Criminology, 61*(4), 965–984. https://doi.org/10.1093/bjc/azaa109

Goyes, D. R., South, N., Sollund, R., & de Carvalho, S. (2021). Editors' introduction to the special issue, "Southern criminologies: Methods, theories and indigenous issues". *Critical Criminology*, *29*(3), 423–429. https://doi.org/10.1007/s10612-021-09586-w

Halsey, M. (2004). Against 'green' criminology. *The British Journal of Criminology, 44*(6), 833–853. https://doi.org/10.1093/bjc/azh068

Hauser, M. (2021). *Mapping water in Dominica: Enslavement and environment under colonialism.* University of Washington Press. https://library.oapen.org/handle/20.500.12657/47757

Hoag, C. (2019). "Water is a gift that destroys": Making a national natural resource in Lesotho. *Economic Anthropology, 6*(2), 183–194. https://doi.org/10.1002/sea2.12149

Hyam, R. (2010). *Understanding the British Empire*. Cambridge University Press.

Jimani, H. (2021, September 5). Police investigates Sayama Tea Estate fracas [News]. *Zodiak.* https://www.zodiakmalawi.com/nw/national-news/66-news-in-southern-region/3825-police-investigates-sayama-tea-estate-fracas

Kalinga, O. (1996). Résistance, politics of protest, and mass nationalism in colonial Malawi, 1950–1960. A Reconsideration. *Cahiers D'études Africaines, 36*(143), 443–454. https://doi.org/10.3406/cea.1996.1424

Kishindo, P., & Mvula, P. (2017). Malawi's land problem and potential for rural conflict. *Journal of Contemporary African Studies, 35*(3), 370–382. https://doi.org/10.1080/02589001.2017.1342784

Lynch, M. J. (2020). Green criminology and environmental crime: Criminology that matters in the age of global ecological collapse. *Journal of White Collar and Corporate Crime, 1*(1), 50–61. https://doi.org/10.1177/2631309X19876930

Malijani, O. (2019). Local stakeholder perspectives on mining development in the Mount Mulanje cultural landscape, Malawi. *Journal of African Cultural Heritage Studies*. https://www.academia.edu/41182754/Local_Stakeholder_Perspectives_on_Mining_Development_in_the_Mount_Mulanje_Cultural_Landscape_Malawi

Mangulama, J. A., & Wu, J. (2022). Land reform or continued social exclusion? Land occupations, state responses and neoliberal policies in Southern Malawi. In F. Mazwi, G. T. Mudimu, & K. Helliker (Eds.), *Capital penetration and the peasantry in southern and eastern Africa: Neoliberal restructuring* (pp. 101–116). Springer International Publishing. https://doi.org/10.1007/978-3-030-89824-3_5

Mathur, C., & Mulwafu, W. (2018). Colonialism and its legacies, as reflected in water, incorporating a view from Malawi. *Wires Water, 5*(4), e1287. https://doi.org/10.1002/wat2.1287

McCracken, J. (2008). *Politics and Christianity in Malawi, 1875–1940: The impact of the Livingstonia Mission in the northern province*. African Books Collective.

McCracken, J. (2012). *A history of Malawi, 1859–1966*. Boydell & Brewer Ltd.

Morris, B. (2016). *An environmental history of Southern Malawi: Land and people of the Shire Highlands*. Palgrave Macmillan.

Mtilatila, L., Bronstert, A., Shrestha, P., Kadewere, P., & Vormoor, K. (2020). Susceptibility of water resources and hydropower production to climate change in the tropics: The case of Lake Malawi and Shire River Basins, SE Africa. *Hydrology, 7*(3), 54. https://doi.org/10.3390/hydrology7030054

Mulwafu, W. (2011). *Conservation song: A history of peasant-state relations and the environment in Malawi, 1860—2000*. The White Horse Press.

Namangale, F. (2021, August 8). Govt seeks to buy tea land. *Nation on Sunday*, 1–4.

Namusanya, D., Rogers, A., & Gilmour, D. (2022). Taking from the rural to serve the urban: The Likhubula water project and the slow violence of water abstraction in Malawi. *Criminological Encounters, 5*(1). https://www.criminologicalencounters.org/index.php/crimenc/article/view/98

Nangoma, D., & Nangoma, E. (2013). *Climate change and adaptation strategies: A case study of the Mulanje Mountain Forest Reserve and its surroundings.* http://www.ndr.mw:8080/xmlui/handle/123456789/641

Ndlovu-Gatsheni, S. J. (2015). Decoloniality as the future of Africa. *History Compass, 13*(10), 485–496. https://doi.org/10.1111/hic3.12264

Nixon, R. (2011). *Slow violence and the environmentalism of the poor*. Harvard University Press.

Orlove, B., & Caton, S. C. (2010). Water sustainability: Anthropological approaches and prospects. *Annual Review of Anthropology, 39*(1), 401–415. https://doi.org/10.1146/annurev.anthro.012809.105045

Pensulo, C. (2018, February 27). Search for new urban water sparks conflict in parched Malawi. *News.Trust.Org*. https://news.trust.org/item/20180227023030-7ml36/

Potts, D. (2012). Land alienation under colonial and white settler governments in southern Africa: Historical land 'grabbing'. In *Handbook of Land and Water Grabs in Africa* (pp. 24–42). Routledge.

Riley, L. (2012). The political economy of urban food security in Blantyre, Malawi. *The African Institute Occasional Paper Series, University of Western Ontar, 1*, 1–16.

Sharma, N. (2022). Commodities trade, river transport and colonialism: The Brahmaputra river valley in the nineteenth century. *The Indian Economic & Social History Review, 59*(1), 75–94. https://doi.org/10.1177/00194646211065355

Sharp, J. (2008). *Geographies of postcolonialism.* SAGE.

Shumba, O. (2011). Commons thinking, ecological intelligence and the ethical and moral framework of Ubuntu: An imperative for sustainable development. *Journal of Media and Communication Studies, 3*(3), 84–96.

Strang, V. (2004). *The meaning of water* (1st ed.). Routledge.

Sultana, F. (2022). The unbearable heaviness of climate coloniality. *Political Geography, 102638*,. https://doi.org/10.1016/j.polgeo.2022.102638

Tchuwa, I. (2017). The 'poisoned chalice' of state ownership of water infrastructure in contemporary Blantyre, Malawi. *Journal of Contemporary African Studies, 36*, 1–22. https://doi.org/10.1080/02589001.2017.1414167

Tchuwa, I. (2019). Colonial remaking of society–nature relations in southern Africa: Reconfiguring subjectivities, producing a commodified waterscape in Blantyre City, 1850s–1890s. *Capitalism Nature Socialism, 30*(3), 70–90. https://doi.org/10.1080/10455752.2018.1446099

White, R. (2021). Global harms and the natural environment. In P. Davies, P. Leighton, & T. Wyatt (Eds.), *The Palgrave handbook of social harm* (pp. 89–114). Springer International Publishing. https://doi.org/10.1007/978-3-030-72408-5_5

Wilson, J. (2018). The effect of deforestation on rainfall in Malawi. *The Society of Malawi Journal, 71*(1), 1–6.

6

Political Economy and the Government Attack on Sharks: A Non-Speciesist Southern Green Criminology

Reece Walters and Amy Couper

Introduction

The esteemed former Justice of the High Court of Australia, and internationally renowned jurist, the Honorable Michael Kirby has for more than a decade predicted that the relationship between humans and non-human animals will become increasingly central to legal jurisprudence, political policy, and public discourse (Kirby, 2009). Indeed, his forecast has been realised with a rapid increase in national and international laws, polices, and frameworks that seek to protect the welfare, and advocate for the rights, of non-human species (LSJ, 2021). For the award-winning

R. Walters (✉)
Deakin University, Melbourne, VIC, Australia
e-mail: reece.walters@deakin.edu.au

A. Couper
Queensland Police Service, Brisbane, QLD, Australia

D. R. Goyes (ed.), *Green Crime in the Global South*, Palgrave Studies in Green Criminology, https://doi.org/10.1007/978-3-031-27754-2_6

and acclaimed philosopher and author of the influential *Animal Liberation*, Peter Singer, humanity will both prosper and become morally and socially enlightened when it sees itself as one with nature, as an equal, not a conqueror (Singer, 2020). However, this sentiment is at odds with international governments of varied political persuasions that continue to pursue anthropocentric policies premised on animal exploitation for economic prosperity (Morton et al., 2020).

The marine species *selachimorpha*, more commonly known as 'sharks', continues to haunt humanity with mystique, danger, and intrigue, and provide a poignant example of government speciesism. There are 370 different species of shark worldwide, yet only a handful pose a threat to humans. Yet, every instance of a 'shark attack' poses an exaggerated media and political response, and creates a visceral human retaliation based on fear, uncertainty, and insecurity. For the most part, sharks are alien to humans and represent an aquatic threat that is feared and fictionalised as a stalking predator that must be captured, tagged, and wherever necessary culled (Zachos, 2019). A common human response to a shark bite, fatal or otherwise, is swift and intolerant, premised on *galeophobic* speciesism and underpinned by fiscal and human safety political priorities. Yet, this is not the experience nor the ideology of Indigenous peoples that believe sharks are an integral part of nature and culture—as creation beings, totems of ancestors, stars of the sea, and controllers of the seasons (Allom, 2015).

This chapter draws on discourses in green and Southern criminology, notably speciesism and political economy, to examine the ways in which nature, in this instance sharks, are socially and politically constructed as a human and economic risk to justify expensive, harmful, and unnecessary marine culling. We utilise a case study approach of an Australian example, drawing on documents obtained under the Freedom of Information Act, together with thousands of public submissions, to critically examine the political economy of a flawed shark mitigation programme. In doing so, we assert that Southern voices aid understandings of nature that seek to avert a 'catch and kill' approach and promote instead human–wildlife harmony. We conclude that non-speciesist discourses from Indigenous societies and previously unheard voices of the global South promote the value of nature in new ways that unearth and critique the failed and flawed commodity and commercially driven government decision-making that promotes the harmful policies that threaten an internationally protected species.

The Political Economy of Shark Bites

There are 500 species of sharks that exist worldwide, yet only 30 are believed to be involved in incidents involving human attacks (ISAF, 2012). The great white shark is disproportionately identified as responsible for the greatest number of bites to humans, yet most are delivered by more common species, such as the bull shark or black tip shark (Naylor, 2022). Yet such smaller species of shark don't attract the media and political attention of that granted to the great white—where the language of 'attack' supplants that of 'bite' in ways underpinned by political economy.

The 1975 Hollywood blockbuster film *Jaws* successfully gripped an international audience with ephemeral fears about a lurking monster in the depths of the oceans where people holiday, an untamed and irrational beast feasting on human swimmers for its survival. Yet this Spielberg classic also conveyed another powerful message: that the great white shark was a threat to local communities, most notably their lifestyles, livelihoods, and seaside trade (Frentz & Rushing, 2009; Robinson, 2017). The controversial Slovenian philosopher, Slavoj Žižek, has argued in his film *The Perverts Guide to Ideology* that the former socialist Cuban leader Fidel Castro, who watched *Jaws*, 'loves the film, once said that for him it was obvious that *Jaws* is kind of a leftist/Marxist film and that the shark is a metaphor for brutal, big, capitalism exploiting ordinary Americans' (Wolters, 2013). There are numerous well cited lines within the movie and none more so than Mayor Vaughn's if you close those beaches, we're finished ... Tomorrow is the 4th of July, and we are going to open for business. It's going to be our best Summer in years. If you're so concerned about the beaches, you two, you do whatever you have to, to keep them safe, but with or without you, those beaches will be open for this weekend (Filmsite, 2020). Thus, the political economy link between humans and sharks, or 'The Jaws Effect', was forever forged and embedded through this iconic film that unmistakably laid bare the competing realities of risk, rights, and responsibilities when the imperatives of local trade clash with nature and issues of human safety, leisure, and consumption (Neff, 2015).

The media connections between shark attacks and the purported decline in local economies are not new (Miller, 1995). Indeed, the dramatic and graphic headlines of press outlets have not only turned people away from beaches and seaside resorts, but have influenced elections, leading some commentators to suggest that shark attacks 'are bad for democracy' (Achen & Bartels, 2016). Yet, as Neff and Hueter (2013) identify, the value-laden reporting that adopts the language of 'man-eater', 'rogue', and 'attack' provide political decision-makers with a discourse of dangerousness that permits the enactment of unwanted or overreaching measures. Sharks are often considered to be amongst the most feared animals in the world, despite the fact that the risk of drownings, rips, surfboard injuries, and stingray and jellyfish encounters are statistically far more likely to result in harm to humans within ocean environments (Curtis et al., 2012). Despite this and the fact that many members of the public support shark conservation, the words 'shark' and 'attack' when put together are still some of the most fear laden, emotionally provocative words (Neff & Hueter, 2013). Increased media coverage and public attention surrounding shark bite incidents tends to place significant pressure on local governments and other relevant agencies to respond to the issue of bather safety (Curtis et al., 2012). Moreover, the way that sharks are portrayed in the media often influences public perception and concomitant government mitigation strategies (Friedrich et al., 2014; Muter et al., 2013).The so-called 'Jaws Effect' refers to the way that fictional representations of sharks in popular culture are used by political actors to explain real life events (Neff & Hueter, 2013). The Jaws Effect causes people to adopt false notions, that sharks, for example, intentionally hunt humans, that human–shark interactions are always fatal, and that the only way to overcome shark related threats is to kill them (Neff, 2015). As Pepin-Neff (2019: 1) concludes:

> policymakers manipulate these highly emotional situations to make it appear like they are protecting the public and doing 'good work'. This is a scam to defraud the public for mere political gain. The real 'sharks' of this story are the politicians.

There is no doubting that shark behaviour and the ways such creatures interact with marine life, as well as humans that enter their habitats, is largely unknown by the beach-going public, yet as a species they are mythologised, politicised, and demonised by those with the power to create and construct images and perceptions for political and profitable gain (Chapman, 2017). This dichotomy of 'politics' and 'truth' has been well established through Foucauldian discourses (see Foucault, 1979; Lotringer, 2007). We will inculcate this into a discussion of how environmental risks or problems—in this case shark attacks—are constructed as both market failures and market opportunities for political profiteering (Mulberg, 2018). As such, political economy is a useful prism to examine the shark–human dynamic as it is 'the kind of discipline one indulges in if one doesn't care about having an easy life. Speaking truth to power, in particular, comes at a hefty price in many countries around the world' (Varoufakis, 2017: 67).

Indigenous Shark Narratives: Not a 'Human Killer' but a Species of Cultural Significance

An essential component of cognitive justice that forms the basis for a Southern green criminological discourse is to listen, learn, and inculcate the unheard voices of the marginalised periphery (Goyes, 2019). The silencing of Indigenous knowledge enables it to be appropriated and legitimated so that the value of such cultural mores passes beyond the control of the originators and is materialised by corporate and government actors that declare sovereignty over it (Brisman et al., 2018). The impacts of global patterns of exploitation of knowledge, and of the legacies of colonialism, are central to Southern and green criminologies. As Amster (2015: 159) argues:

> Resource wars and patterns of economic colonization are often initiated by the nations of the Global North vis-à-vis those of the Global South, yielding a two-tiered world of privileged consumers at the top and vulnerable producers on the bottom. The false security created by such a system is reinforced by a mindset in which human cultures are seen as separate from nature, and where traditional societies that exist closer to nature are viewed (in Social Darwinist terms) as inferior to modern societies in their sociopolitical, economic, and moral development. These dichotomies (North/South, Nature/Culture, Traditional/Modern) are historically untenable, ecological destructive and self-refuting even when taken at face value. But even more problematically, these dualisms often provide the ideological software that serves to perpetuate an unsustainable world in which people are alienated from one another and are dislocated from the essential workings of the environment all at once.

It should be noted that Amster does not specifically identify Southern and green criminologies, yet such words are essential, in our view, for understanding the premise of a criminological project premised on cognitive justice and the democratisation of knowledge. Moreover, the importance of a nature-centred knowledge, based in the experiences and cultures of Indigenous peoples, is imperative for an innovative Southern green criminological enterprise that seeks new ways of understanding, preserving, and sustainably progressing the human/non-human nature complex. Such an approach is crucial for the subject of this chapter, notably for rethinking speciesist notions of sharks (Jacobs, 2005), not as human-stalking predators of the deep, but as time-honoured, respected, majestic, and mystic creators (Goyes, 2015; Goyes et al., 2021). As Allom (2015: 2) poignantly identifies: 'For Aboriginal and Torres Strait Islander people, sharks have an important history. They are creator beings, ancestors, totems. Their lifecycles reflect the seasons, the landscape and sea country. They are seen in the movement of the stars.'

Throughout time, sharks have been feared by sea traders, employers, and colonisers as 'beasts' that terrorised ship-wrecked vessels and decimated cargo and catch. The Indigenous narrative, however, is a far cry from the evil lurking predator and human-hunting monster of the ocean that so often permeates popular and media culture. Rather than constructing sharks as a danger to be feared and destroyed, Indigenous knowledges teach reverence and respect, where the qualities of the shark are to be respected and inculcated into human lifestyles and attitudes (Castro, 2014). In Australia's Indigenous societies, sharks have had a central place in 'the dreaming' or the story-telling of a 40,000-year-old culture (Kenyon, 2018). Moreover, such majestic sea creatures have symbolised 'bravery, fearlessness and uncanny powers' (McIntyre & Dobson, 2017: 1). Within traditional Hawaiian culture, for example, the chiefs and tribal leaders were referred to as 'sharks walking on land' to signify their status as bearers of wisdom, strength, and authority (Archer, 2018: 7). In the Caribbean, sharks have for thousands of years been a source of food and tools, and thus perceived as a life blood of human and aquatic culture to be protected and preserved (Caribbean Shark Coalition, 2021). In Greenland, Inuit communities have honoured sharks for their meat as a food source as well as the oil from their bodies for cooking and warmth (Kuhnlein & Humphries, 2019). In South Pacific cultures, sharks were integral to societal diets and hunting, and instrumental for ceremonial purposes, notably for young men transitioning into adulthood (American Museum of Natural History, 2019). And sharks have long held cultural significance in numerous Indigenous societies around the world including the New Zealand Maori (Asbjørn & Aich, 2015). Such narratives identify that sharks have been an essential part of human history for millennia, as food, energy, and source for tools and weapons. In addition, as keepers of the oceans, and an inspirational symbol of courage, power, and purpose, they have been constructed in many cultures as 'companion creatures' (Harraway, 2016: 6).

Indigenous knowledges of sharks are not however inculcated into the government policies of highly industrialized and fiscally advanced societies that are dominated by knowledges of the global North where sharks are perceived as an impediment to fiscal prosperity. Within Indigenous societies sharks are 'embraced' through catch, in a way where the creature is respected, honoured, and consumed, but always within the cultural confines of sustainability and mutual reverence. It is estimated that more than 100 million sharks are killed each year for sport, food, and oil in both legal and illegal fishing activities (McCallum & Pan, 2021). The exploitation of sharks for economic prosperity is a global industry estimated to be worth near US$1 billion a year (Van Houtan et al., 2020). In addition to sharks being killed for global trade, they are also culled to enhance human safety and security, notably at coastal economies that profit from beachgoing leisure and livelihoods. The following section exemplifies this point and provides a case study of how speciesism and fear determined an unwarranted and failed shark mitigation strategy in Australia. It is an example of how government policy, driven by fiscal and political imperatives, fails to embrace the knowledges of the marginalised global South, and in doing so alienates Indigenous expertise and the voices of informed and impassioned local communities.

Case Study: 'Catch and Kill'—The Western Australia Shark Hazard Mitigation Drum Line Program (SHMDLP)

Throughout the last two decades, Western Australia (WA) has had a relatively high international rate of shark bites to humans (Couper & Walters, 2021). Almost a decade ago, it was announced that, where previously the government had targeted large white sharks suspected of being responsible for bites, it would pre-emptively catch and destroy all large white, tiger, and bull sharks in the areas around popular swimming beaches, otherwise referred to as 'shark kill zones' (Department of Premier and Cabinet, 2014: 11). The government decision to implement shark drum lines off the WA coastline received a significant amount of public and media attention. This included two protest rallies, drawing crowds of up to 6000 people, as well as protests in South Australia, Queensland, New South Wales, Victoria, and New Zealand (Australian Associated Press, 2014). Regardless of unprecedented public opposition, the catch and kill policy of the WA government went ahead in the name of public safety, business, and tourism. This was the beginning of what was popularly referred to as the 'WA catch and kill' policy. This research sought to understand the background and political nuances of government policy that actively targeted and destroyed a protected marine species.

The first phase of this research accessed publicly available policy and discussion papers from the government, including various state departments, the Federal Department of Environment, and local governments in WA, as well as the tourism industry, scientific community, the public, environmental activists, and the media. In addition to publicly available documents, relevant Hansard excerpts were chronicled, as well as government media releases and official policy documents. A total of 16 separate Freedom of Information (FOI) requests provided a rich and nuanced set of 2000 pages of communications between government decision-makers and relevant policy administrators. The second phase of the research comprehensively examined over 12,000 public submissions made to the Western Australia Environmental Protection Authority in response to the government's recommendation to continue the operation of the WA SHMDLP into the summers of 2014–2017.

In sum, this research identified that the media used emotive language to frame shark bite events, which heightened public anxiety. However, despite this increase in anxiety, there was no association between culling and public support for the WA government's shark kill policy (cf. McCagh et al., 2015). This research examined considerable media coverage of the shark attacks including 690 articles from 94 publications, including the Australian Associated Press, Australian Broadcasting Corporation, *The Sunday Times*, *The Australian*, *The Times*, *The West Australian*, and Agence France Presse. Within the dataset, 303 articles specifically referred to the shark bite incidents as a social problem. For example, the media repeatedly asserted three common themes: endangerment, loss, and entitlement. Illustrating the issues, six articles emphasised the ongoing danger with headlines such as 'Shark attack stirs fears of rogue man-eater stalking Australian coast'. This powerful rhetoric was combined with notions of loss. In 62 articles the message of surfers giving up their pastime due to fears of shark bites was projected constantly into the public psyche, with statements like: 'I've surfed since I was five and it's something I love so much, but it's no longer going to be a part of my life'. This was coupled with a sense of human entitlement over ocean use with headlines like: 'Populated swimming and surfing areas are not "their territory", they are ours. We have the same right to be there as any other species'. These recurring themes were bolstered by using rhetorical motifs to reinforce notions of danger, loss, and entitlement. For example, 'shark attack', 'deadliest place in the world', and 'shark attack capital' appeared in almost 95% of all articles and was adopted by officials in public and political debate. A further 51 articles identified 'the increasing population of sharks', the 'concerns for human safety', and the 'impacts on tourism'—all unproven and unsubstantiated with scientific and reliable evidence (Australian Shark Attack File, 2020; CSIRO, 2020).

Documents obtained through FOI demonstrate that Tourism WA did not have any data on this issue, indeed the official view was that tourism was not impacted in any negative way (Couper & Walters, 2021). Of considerable note is the repeated reporting of 'kill', 'cull', and 'nets' as the pressing and urgent response required to safeguard beachgoers. Specific parliamentary questions were raised by opposition members requesting scientific evidence. The questions asked were: 'Will the minister please list all the sources of scientific research that the DF is aware of to show that the deployment of drum lines off WA will reduce attacks on humans by white sharks? Is the minister aware that the target species of sharks for drum lines in QLD is bull sharks? Is the minister aware that "most shark scientists" are doubtful that drum lining will be effective at reducing fatal attacks on humans by white sharks?' (Couper & Walters, 2021).

While the Minister for Fisheries requested the questions be answered during a recess, it is noted that a lack of scientific evidence for drum lining effectiveness was a recurring criticism of the WA SHMDLP and that the Minister for Fisheries was aware of this prior to its ultimate implementation. Despite this, the WA Premier insisted that the programme was supported by the 'silent majority'; the Western Australian Government proceeded to implement drum lines in January 2014. However, the public submissions collected under an FOI suggest that WA political leader claims were not supported by public opinion and were indeed fabricated (Couper & Walters, 2021).

The WA Environment Protection Authority held a ten-day public submission period in which community members were able to comment on whether it should reassess and rescind the shark catch-and-kill-policy. A total of 628 supported and submitted the following position:

> Sharks that maintain humanities life support systems are in big trouble globally and what exists off the WA coast is rare and unique and is of global significance and importance and must be protected.

A substantial number of public submissions (1509) also referred to 'alternative measures' to catch and kill, including education campaigns, increased aerial patrols, increased beach signage, further research into shark behaviour, and catch and release drum lining. In sum, 60% opposed the proposed policy on animal welfare and environmental grounds and the majority felt it would be ineffective (Couper & Walters, 2021). Regardless of public and political opposition, the WA Government moved forward with a mitigation method that killed a species protected under international law, as well as inadvertently decimating a variety of marine species caught in nets and drum lines.

The Program intended to target white sharks, which is a species listed as 'vulnerable' by the International Union for the Conservation of Nature (IUCN) Red List, in Appendix II of the international Convention on International Trade in Endangered Species (CITES) Treaty, and Appendices I and II of the Convention of Migratory Species (CMS). In Australia's effort to uphold their commitments to these agreements, white sharks are a protected species under the Environment Protection and Biodiversity Conservation (EPBC) Act 1999 (Cth), which outlaws the killing, injuring, taking, trading, keeping, or moving of this species without a permit (Legislative Council, 2013). However, exemptions from Australian legislation may be permitted under extenuating circumstances. Under section 158 of the EPBC Act 1999 (Cth), persons may apply to the Minister for the Environment 'for an exemption from a specified Provision' and the Minister may approve this submission 'only if he or she is satisfied that it is in the national interest that the provisions not apply in relation to the person or the action'.

In January 2014, the Federal Minister for Environment sent a letter to the Premier granting an exemption from the EPBC Act 1999. It was concluded by the former conservative LNP Commonwealth Government (voted out in May 2022) that evidence provided by the WA State Government demonstrated a clear statistical increase in shark bite events in Western Australia to the point where it was determined to be within the national interest to have an exemption under the EPBC Act granted for public safety and economic reasons. Hence, the protections afforded to sharks under federal legislation were waived for unfounded economic and public interests (Couper & Walters, 2021; Neff & Wynter, 2018).

Without scientific evidence, without public support, and without an economic risk assessment, the WA Government went ahead and implemented a kill policy of an internationally protected species based on a media-fuelled knee-jerk approach that highlighted and exaggerated imminent threats to human safety and falsely forecasted devastating losses to local economies. The policy to catch and kill sharks was, therefore, premised on a lie, a fiction, a media generated moral panic that political actors perceived would appeal to an anxious public whilst artificially bolstering the public safety and economic self-interests of elected officials.

Discussion: A Southern Green Non-Speciesist Criminology

The emergence and evolution of Southern and green criminology as dissenting reactions against the status quo in criminology provide standpoints from which to reconsider the contemporary causes and distribution of various forms of injustice, exploitation, and harm (Brisman et al., 2018). A non-speciesist criminology has long highlighted non-human abuse and advocated for both animal welfare and rights (Beirne, 1999; Cazaux, 1999; Regan, 2000). The combination of Southern, green, and non-speciesist criminology, we argue, provides a powerful trilogy to further the conservation of protected species whilst advancing the importance of the human–non-human relationship alluded to in the introduction. In doing so, we develop Sollund's (2021) argument for a green critical non-speciesist criminology by engaging in new speciesism debates.

For some animal studies commentators, speciesism debates polarised by principles of 'equal consideration of interests' and 'moral relativism' (see Kagan, 2016; Singer, 2016) are capable of moving forward through 'integrating empirical research with philosophical analysis' (Hopster, 2019). The 'Singer–Kagan dispute' and their respective views based on notions of prejudice remain widely debated (Director, 2020) and our intention here is not to rehearse or critique the nuances of those ongoing debates. Instead, we propose to enliven the debate with empirical examples, through the case study discussed above. Our critique of the WA SHMDLP draws on new speciesist philosophers such as Steven Wise who advocates for a flawed hierarchy of animals with humans at the apex. In our view, it is this thinking that determined and shaped government decision-making in the case study discussed above.

According to Wise (2000), new speciesism arguments assert that animals are ranked as more or less deserving of moral consideration based on a range of factors (Wyatt, 2013). For example, humans tend to assign animals differing levels of value based on their genetic closeness to us or the roles they play in our lives (be it companion animals or those used for the production of meat) (Jean Moore, 2013: 12). This means that humans, notably those of the industrialised global North, are more likely to value the lives of primates and other mammals, especially companion animals, over rodents or other vermin and creatures genetically different from humans, such as insects, fish, or reptiles. Indeed, Bekoff (2013: 13) argues that other great apes and primates, followed by companion animals, are likely to be viewed as being 'higher, better, or more valuable' than birds or fish. Fish provide an example of a taxonomic class of animals that are frequently subjected to new speciesism, despite scientific evidence that, as sentient beings, they have long-term memories and have demonstrated sophisticated cognitive behaviours (Bekoff, 2013; Braithwaite, 2010). This might be attributed to the debate around whether fish can feel pain. While some studies suggest that they do have the capacity to feel pain, others suggest that they do not (Key, 2016; Rose, 2016). Furthermore, species within this class are also attributed different levels of value. A species of fish that is iconic, for example, one used as a charismatic species in a conservation campaign, is likely to be attributed more value than a lesser-known species or one that is not facing extinction. Charismatic species refer to species selected to represent conservation campaigns and are 'species with popular appeal and that tend to attract wide support for protection against environmental change' (Spellerberg, 2014). Species used for this purpose tend to be selected on a range of traits that they possess, including how rare, beautiful, impressive, endangered, cute, or dangerous they are perceived to be (Albert et al., 2018). Well known examples of charismatic species include tigers, polar bears, and orang-utans, while fish are routinely subjected to new speciesism. While Albert et al. (2018) use the term 'charismatic species' to describe these animals, they recognise that most of these species are large mammals, which is consistent with the term 'charismatic megafauna' that is also used to describe these animals. Hund (2012) concurs that subjective qualities like cuteness or cuddliness contribute to the appeal of these species but also suggests that the large size of these animals plays a role in the value that they are attributed. Therefore, an elephant is more likely to stimulate support for conservation programmes than an equally endangered species of mouse, despite both being mammals.

In considering the role of new speciesism in the marine environment, dolphins and other cetaceans present challenges to those who subscribe to new speciesism. While they are more genetically similar to humans, as mammals, than fish or birds, and have been scientifically proven to be highly intelligent and socially complex creatures, they behave in a way that is less 'human' and more 'fish' than many other mammals (Marino, 2013). That is, they move differently from humans, do not have clearly recognisable facial expressions, and live in a very different physical environment than, for example, apes or other terrestrial mammals. Despite this, in modern cultures of the global North, dolphins are perceived favourably by humans and to be deserving of protection and conservation (Marino, 2013). Indeed, dolphins are also listed by Albert et al. (2018) as one of the world's top 20 most charismatic species, who argue that they are rare, beautiful, impressive, endangered, and cute, but not dangerous.

Another group of marine animals that challenge the typical hierarchy of animals according to new speciesism (where mammals > reptiles) are sea turtles. While the majority of the world's most iconic charismatic species are mammals (Albert et al., 2018), sea turtles have become recognised as a conservation icon for the marine environment, whereby their position on the human moral compass has been increased from that of most reptiles. It has even been argued that sea turtles have displaced whales as charismatic species used to promote the environmental movement that aims to conserve the marine environment (Rieser, 2014). Overall, the literature suggests that while humans inherently attribute different levels of value to animals based on their genetic closeness to us, other factors, particularly whether they represent conservation icons as charismatic species, can undermine this basic structure and lead to scenarios where, for example, a fish or reptile is considered more valuable than a mammal.

In the case study discussed above, the great white shark was not framed as a charismatic, endearing creature with human-like characteristics but as an apex predator, a killer, and an imminent danger to public safety and security. Its meaning as a species protected under international law was not constructed by its intrinsic and dynamic value as an integral part of marine ecosystems with crucial cultural significance, but rather as a monster, a threat to humanity—the demonised animal other (Sollund, 2017). As such, government actions that cull and kill nonhuman creatures, purely for unscientific and unproven perceived threats to people and pocket, are from a Southern green non-speciesist criminological position complicit in acts of theriocide—the unnecessary and/or unlawful killing of animals by humans (Beirne, 2014)—and contradict the visions of most original inhabitants of the territories, who hold sharks in high regard.

Conclusion

We conclude that non-speciesist discourses from Indigenous societies and previously unheard voices of the global South promote the value of nature in new ways that unearth and critique the failed and flawed commodity and commercially driven government decision-making policies that promote the harmful catch-and-kill policies of an internationally protected species and the marine destruction associated with it.

We conclude that the media falsely constructed and inflated 'shark attacks' in Western Australian as a menacing and dangerous social problem of significant public concern. The media frequently appeared to represent public opinion; however, the examination of public submissions to the Western Australia Environmental Protection Authority determined that the public was predominantly opposed to drum lining, and that, in our view, the media grossly misrepresented public interests. Moreover, we conclude that media reporting of shark attacks in Western Australian during the period under review favourably represented the interests of the Western Australian tourism industry and the economic priorities of the state government. Yet there is no evidence that Western Australian state tourism suffered as a result of shark attacks. The then Liberal-National Government was initially reluctant to implement lethal shark hazard mitigation techniques, but was persuaded by media misrepresentation of public opinion and the economic priorities of the conservative Federal Australian Government.

We conclude that sharks and shark attacks are exploited as both objects and foci for political and economic manipulation. Irrespective of their overwhelming historical and contemporary non-fatal interactions with humans, the reported isolated bites and fatalities are manipulated and capitalised on by political actors for economic means. The great white shark, with the assistance of popular culture that often emphasises aggression and human tragedy, has the power to evoke imagery and emotions of fear, danger, and risk. Such 'othering' and demonisation of nature provides fertile ground for political constructivist coercion. A position not supported by either scientific evidence nor public opinion, nevertheless often serves to endorse and justify catch-and-kill shark policies, in breach of international law.

This research also concludes that the public's perception of sharks, contrary to notions of new speciesism, is positive and that there is an increasing level of awareness about the conservation efforts required to maintain these species at healthy population levels. The public, in

Western Australia during the time of this research, were overwhelmingly unsupportive of lethal shark hazard mitigation techniques implemented in response to shark bite events. However, state and federal political actors capitalised on atypical events to construct a false narrative that reified a response premised on economic priorities and not the social and cultural well-being of those citizens most intimately affected.

Finally, we conclude that Indigenous knowledges are central to the conservation of protected wildlife. When combined with Southern and green criminological perspectives, such narratives have the powerful potential to shape and influence future government environmental policies. This chapter, through a Southern green non-speciesist perspective, offers a useful contribution to the literature on public perceptions regarding marine policy, where wildlife has the potential to be negatively impacted, by helping to inform policy makers responsible for developing marine policies, particularly those designed to enhance public safety.

References

Achen, C., & Bartles, L. (2016). *Blind retrospection: Why shark attacks are bad for democracy.* Centre for the Study of Democratic Institutions. https://www.vanderbilt.edu/csdi/research/CSDI_WP_05-2013.pdf. Accessed 14 Apr 2020.

Albert, C., Luque, G., & Courchamp, F. (2018). The twenty most charismatic species. *PLoS ONE, 13*(7), 1–12. https://doi.org/10.1371/journal.pone.0199149

Allom, L. (2015). Indigenous cultural views of the shark. https://www.abc.net.au/radionational/programs/earshot/indigenous-cultural-views-of-the-shark/6798174

American Museum of Natural History. (2019). Using shark-teeth tools to track changes in biodiversity. https://www.amnh.org/explore/news-blogs/on-exhibit-posts/shark-tooth-weapons-track-biodiversity

Amster, R. (2015). *Peace ecology*. Paradigm Publishers.

Archer, S. (2018). Sharks upon the land. Colonialism, indigenous health and culture Hawaii 1788–1855. Cambridge University Press.

Australian Associated Press. (2014, January 2). Shark cull: Nationwide protests against 'catch and kill' policy in WA. *The Guardian*. https://www.theguardian.com/world/2014/feb/01/thousands-protest-against-wa-shark-cull

Australian Shark Attack File. (2020). https://taronga.org.au/conservation-and-science/australian-shark-attack-file

Beirne, P. (1999). For a nonspeciest criminology: Animal abuse as an object of study. *Criminology: An Interdisciplinary Journal, 37*(1), 117–149.

Beirne, P. (2014). Theriocide: Naming animal killing. *International Journal of Crime, Justice and Social Democracy, 3*(2), 50–67.

Bekoff, M. (2013). Who lives, who dies, and why? How speciesism undermines compassionate conservation and social justice. In *The politics of species: Reshaping our relationships with other animals*. Cambridge University Press.

Braithwaite, V. (2010). *Do fish feel pain?* Oxford University Press.

Brisman, A., South, N., & Walters, R. (2018). Southernizing green criminology: Human dislocation, environmental injustice and climate apartheid. *Justice, Power and Resistance, 2*(1), 1–21.

Caribbean Shark Coalition. (2021). Building capacity for the conversation of sharks and rays in the Caribbean. https://caribbeansharks.co/

Castro, J. (2014). Historical knowledge of sharks: Ancient science, earliest American encounters and American science, fisheries, and utilization. *Marine Fisheries Review, 74*(4), 1–26.

Cazaux, G. (1999). Beauty and the beast: Animal abuse from a non-speciesist criminological perspective. *Crime, Law and Social Change, 31*, 105–125.

Chapman, B. (2017). *Shark attacks: Myths, misunderstandings and human fear*. CSIRO Publishing.

Couper, A., & Walters, R. (2021). The great white bite: A critique of the Western Australian government's shark hazard mitigation drum line program. *Journal of Sociology*. https://doi.org/10.1177/1440783320964556

CSIRO. (2020). White shark research findings. https://www.csiro.au/en/Research/Environment/Oceans-and-coasts/Sharks/White-shark-research-findings

Curtis, T. H., Bruce, B. D., Cliff, G., Dudley, S. F. J., Klimley, P. A., Kock, A. A., Lea, R. N., Lowe, C. G., McCosker, J. E., Skomal, G. B., Werry, J. M., & West, J. G. (2012). Responding to the risk of white shark attack: Updated statistics, prevention, control methods and recommendations. In *Global perspectives on the biology and life history of white shark* (pp. 477–509). CRC Press.

de Sousa, B. (Ed.). (2007). *Cognitive justice in a global world: Prudent knowledges for a decent life*. Lexington.

Department of Premier and Cabinet. (2014). *Review: Western Australia shark hazard mitigation drum line program 2013–2014*. Government of Western Australia. https://www.dpc.wa.gov.au/Publications/Documents/Review%20-%20Western%20Australia%20Shark%20Hazard%20Mitigation%20Drum%20Line%20Program%202013-14.pdf

Director, S. (2020). Speciesism, prejudice and epistemic peer disagreement. *The Journal of Value Inquiry.* https://doi.org/10.1007/s10790-020-09735-4

Filmsite. (2020). JAWS 1975. *Filmsite Movie Review*. https://www.filmsite.org/jaws.html

Foucault, M. (1979). Birth of biopolitics (Michel Foucault: Lectures at the College De France). Palgrave.

Frentz, T., & Rushing, J. (2009). Integrating ideology and archetype in rhetorical criticism, part II: A case study of Jaws. *Quarterly Journal of Speech, 79*(1), 61–81.

Friedrich, L. A., Jefferson, R., & Glegg, G. (2014). Public perceptions of sharks: Gathering support for shark conservation. *Marine Policy, 47*, 1–7. https://doi.org/10.1016/j.marpol.2014.02.003

Goyes, D. (2015). Denying the harms of animal abductions for biomedical research. In R. Sollund (Ed.), *Green harms and crimes: Critical criminology in a changing world* (pp. 170–188). Ashgate.

Goyes, D. (2019). *Southern green criminology: A science to end ecological discrimination.* Emerald.

Goyes, D., Abaibira, M. A., Baicué, P., Cuchimba, A., Ñeñetofe, D. T. R., Sollund, R., South, N., & Wyatt, T. (2021). Southern green cultural criminology and environmental crime prevention: Representations of nature within four Colombian indigenous communities. *Critical Criminology, 29*(3), 469–485. https://doi.org/10.1007/s10612-021-09582-0

Haraway, D. (2016). *Staying with the trouble: Making Kin in the Chthulucene.* Duke University Press.

Hopster, J. (2019). The speciesism debate: Intuition, method and empirical advances. *Animals, 9*(2). https://www.ncbi.nlm.nih.gov/pmc/articles/PMC6940905/

Hund, A. (2012). Charismatic megafauna. In *Encyclopedia of global warming & climate change*. SAGE.

International Shark Attack File. (2022). Species implicated in attacks. https://www.floridamuseum.ufl.edu/shark-attacks/factors/species-implicated/

Jacobs, G. (2005). Speciesism. *Anthrozoos, 18*(4), 443–445. https://doi.org/10.2752/089279305785594018

Jean Moore, L. (2013). Speciesism. *Contexts, 21*(1), 12–13. https://doi.org/10.1177/1536504213476240

Jøn, A. A., & Aich, R. S. (2015). Australian Folklore 30, 2015 Australian Folklore 28, 2013 Southern Shark Lore forty years after Jaws: The positioning of sharks within Murihiku, New Zealand.

Kagan, S. (2016). What's wrong with speciesism? *Journal of Applied Philosophy, 33*, 1–21. https://doi.org/10.1111/japp.12164

Kenyon, G. (2018). Australia's ancient language shaped by sharks. https://www.bbc.com/travel/article/20180429-australias-ancient-language-shaped-by-sharks

Key, B. (2016). Why fish do not feel pain. *Animal Sentience, 1*(3), 1.

Kirby, M. (2009). *Animal law lecture series*. Voiceless. https://voiceless.org.au/animal-law/lecture-series/

Kuhnlein, H., & Humphries, M. (2019). Traditional animal foods of indigenous peoples of northern North America—The contributions of wildlife diversity to the subsistence and nutrition of indigenous cultures. http://traditionalanimalfoods.org/fish/saltwater/page.aspx?id=6440

Legislative Council. (2013). *Extract from Hansard*. Government of Western Australia. http://www.parliament.wa.gov.au/Hansard/hansard.nsf/0/8ae7ed9f119a731348257c4000218afc/$FILE/C39+S1+20131211+p7596d-7596d.pdf

Lotinger, S. (2007). *The politics of truth*. MIT Press.

LSJ. (2021). Animal law update: Positive changes for animals in 2021. https://lsj.com.au/articles/animal-law-update-positive-changes-for-animals-in-2021/

Marino, L. (2013). Humans, dolphins, and moral inclusivity. In *The politics of species reshaping our relationships with other animals* (pp. 95–105). Cambridge University Press.

McCagh, C., Sneddon, J., & Blanche, D. (2015). Killing sharks: The media's role in public and political response to fatal human–shark interactions. *Marine Policy, 62*, 271–278.

McCallum, W., & Pan, W. (2021). 100 million dead sharks—It's not all about shark fin soup. https://www.greenpeace.org/international/story/46967/100-million-dead-sharks-its-not-all-about-shark-fin-soup/

McIntyre, K., & Dobson, B. (2017). The shark in Nyungar culture. https://anthropologyfromtheshed.com/project/shark-in-nyungar-culture/

Miller, S. (1995). Sharks take a bite out of seaside economy. *New Scientist, 147*, 1–10.

Morton, R., Hebart, M., & Whittaker, A. (2020). Explaining the gap between the ambitious goals and practical reality of animal welfare law enforcement: A review of the enforcement gap in Australia. *Animals, 10*(3), 482.

Mulberg, J. (2018). The challenge of environmental governance: Ecology and the need for a heterodox political economy. *Journal of Australian Political Economy, 80*, 129–154.

Muter, B. A., Gore, M. L., Gledhill, K. S., Lamont, C., & Huveneers, C. (2013). Australian and US news media portrayal of sharks and their conservation. *Conservation Biology, 27*, 187–196.

Naylor, G. (2022). Yearly worldwide shark attach summary. https://www.floridamuseum.ufl.edu/shark-attacks/yearly-worldwide-summary/

Neff, C. (2015). The Jaws effect: How movie narratives are used to influence policy responses to shark bites in Western Australia. *Australian Journal of Political Science, 50*, 114–127.

Neff, C., & Hueter, R. (2013). Science, policy, and the public discourse of shark "attack": A proposal for reclassifying human–shark interactions. *Journal of Environmental Studies and Sciences, 3*, 65–73.

Neff, C., & Wynter, T. (2018). Shark bites and shark conservation: An analysis of human attitudes following shark bite incidents in two locations in Australia. *Conservation Letters, 11*, 1–8.

Pepin-Neff, C. (2019). *Flaws. Shark bites and emotional policymaking*. Palgrave.

Regan, T. (2000). *Defending animal rights*. University of Illinois Press.

Rieser, A. (2014). *The case of the green turtle: An uncensored history of a conservation icon*. John Hopkins University Press.

Robinson, D. (2017). Monster or great fish? Peter Benchley's Jaws as ecocritical text. *Scrutiny2, 21*(3), 57–72. https://doi.org/10.1080/18125441.2016.1255246

Rose, J. (2016). Pain in fish: Weighing the evidence. *Animal Sentience, 3*(25).

Singer, P. (2016). Why speciesism is wrong: A response to Kagan. *Journal of Applied Philosophy, 33*, 31–35. https://doi.org/10.1111/japp.12165

Singer, P. (2020). *Why vegan*. Penguin.

Sollund, R. (2017). The animal other: Legal and illegal theriocide. In M. Hall, J. Maher, A. Nurse, G. Potter, N. South, & T. Wyatt (Eds.), *Greening criminology in the 21st century. Contemporary debates and future directions in the study of environmental harm*. Routledge.

Sollund, R. (2021). Green criminology: Its foundation in critical criminology and the way forward. *The Howard Journal of Crime and Justice, 60*(3), 304–322.

Spellerberg, I. F. (2014). Charismatic species. In *Encyclopedia of environmental change* (pp. 150–152). SAGE.

Van Houtan, K. S., Gagné, T. O., Reygondeau, G., Tanaka, K. R., Palumbi, S. R., & Jorgensen, S. J. (2020, October 28). Coastal sharks supply the global shark fin trade. *Biology Letters, 16*(10), 20200609. https://doi.org/10.1098/rsbl.2020.0609

Varoufakis, Y. (2017). Taking the red or blue pill? *Journal of Australian Political Economy, 80*, 65–73.

Wise, S. (2000). *Rattling the cage: Toward legal rights for animals.* Hachette Books.

Wolters, E. (2013). Zizek on jaws and fascism. http://www.critical-theory.com/watch-zizek-jaws-fascism/

Wyatt, T. (2013). *Wildlife trafficking, a deconstruction of the crime, the victims and the offenders.* Palgrave Macmillan.

Zachos, E. (2019). Why are we afraid of sharks? There's a scientific explanation. *National Geographic.* https://www.nationalgeographic.co.uk/animals/2019/06/why-are-we-afraid-of-sharks-theres-a-scientific-explanation

Part II

Responses to Environmental Crime in the Global South

7

Green Potential in the Global South: The *Phulbari Movement* in Neoliberal Bangladesh

Nikhil Deb and Avijit Chakrabarty Ayon

Introduction

In this chapter, we analyse the *Phulbari* Movement—the most significant and longest anti-mining resistance movement in Bangladesh—as an example of a distinct type of justice struggle in a remote region of the global South. We specifically examine the ways in which the *Phulbari* Movement effectively resisted market forces trying to destroy the life, livelihood, and environment of local people. The proposed Phulbari coal mining project in the Dinajpur district of Bangladesh would have affected Indigenous communities, dispossessed hundreds of thousands, and harmed vast water resources and thousands of acres of

N. Deb (✉)
California Polytechnic State University, San Luis Obispo, California, US
e-mail: ndeb@calpoly.edu

A. C. Ayon
Shahjalal University of Science and Technology, Sylhet, Bangladesh

D. R. Goyes (ed.), *Green Crime in the Global South*, Palgrave Studies in Green Criminology, https://doi.org/10.1007/978-3-031-27754-2_7

fertile agricultural land in the area. We draw on existing research on dispossession and environmental justice to highlight the importance of studying both issues in tandem, rather than shedding light on one or the other. We propose that bringing together land dispossession and environmental justice scholarship, as evident in the objectives and success of the *Phulbari* Movement, advances our understanding of the factors forcing marginalized communities to become the influential actors of resistance to neoliberal development projects in the global South.

The Case and Objectives

BHP Billiton, an Australia-based mining company, ill-reputed for owning and operating the OK Tedi mine, discovered coal in the Phulbari region of Bangladesh in the mid-1990s. In 2005, BHP Billiton, considering the area's vast groundwater reserve, sold its mining rights to Asia Energy Corporation Private Limited (Asia Energy), a subsidiary of London-based coal mining company GCM Resources. The proposed mining area comprised more than 100 villages. The project would have dispossessed hundreds of thousands over the mine's lifespan, affected acres of fertile agricultural land, and polluted the surrounding environment, including water resources (Bedi, 2015; Hasan, 2020; Luthfa, 2011). Despite such impending consequences, the Bangladeshi government and the corporation continued to pursue the project in this rich agroecological zone. However, local activists presented a forceful challenge against the top-down narrative of economic growth, highlighting this region's critical social and ecological dimensions.

Because of its critical location near rivers and on vast agricultural lands, the proposed project in Phulbari would have devastated the region's agricultural, economic, and environmental health. However, in its feasibility studies, Asia Energy totally discounted many crucial aspects of the local economy, including severe damage to the environment and loss of agricultural land and fertility. Additionally, Phulbari is also home to several Indigenous peoples. Constructing an environmentally destructive coal plant in this region is another example of corporations targeting lands on which Indigenous and minority people live (Kirsch,

2014; Pellow, 2018; Schlosberg & Carruthers, 2010). Scholars show that marginalized rural inhabitants and Indigenous peoples have been victims of "land grabbing" since the country's independence (Chowdhury, 2018).

As market liberalization in Bangladesh bolstered in the 1990s (World Bank, 1995), privatizing a massive number of lands further propelled land grabbing in the country (see Adnan, 2013; Bedi, 2015; Deb, 2020b; Feldman and Geisler, 2012; Paprocki, 2021). It should be noted that massive policy reform in the energy sector supporting the rules of the Washington Consensus was evident in Bangladesh in the early 1990s. Among these policy changes is the production sharing contracts (PSCs) with several international corporations in the mid-1990s, when the role of local agencies was substantially decreased (Deb, 2020b; Muhammad, 2014; Pegu, 2011). The Phulbari coal mining project, as we demonstrate, was one of many examples of how corporations and states attempt to invalidate the demands and aspirations of local people.

Emphasizing the economic potential and framing it as a "landmark foreign investment project" (Chowdhury, 2019; Faruque, 2018a) to solve the energy problem in Bangladesh, Asia Energy (2006) reported that the proposed project, although it would cause substantial socio-environmental damage according to the report, had overwhelming support from local communities. Such bold gestures from corporations, despite major resistance on the ground, are not uncommon in countries of the global South. The reality on the ground in Phulbari was different from what Asia Energy reported. The local communities and several activist organizations protested against the coal mine project as early as March 2005. They documented that the Phulbari coal mine would cause massive loss of agricultural land and of diverse crops over the span of the mine's life.

On the other hand, Asia Energy and the Bangladeshi government attempted to persuade people to support the project (Hasan, 2020; Luthfa, 2011). However, even in the face of manipulation and torture and police-military attacks, the activists refused to bow to power, continuing the resistance for over a decade. Despite repeated attempts by the state and corporation to distract, suppress, and stop the resistance, the local communities remained determined to prevent the corporate destruction of their society, land, and the environment.

The *Phulbari Movement* is the first resistance movement in Bangladesh that successfully confronted a project that would have caused dispossession and environmental destruction in this marginalized region. Drawing on dispossession and environmental justice scholarship and analysing data from documents, reports, published interviews, and the first author's direct observation of several movement activities, we present and examine the *Phulbari Movement* as an anti-dispossession environmental justice movement in the global South. The chapter proceeds by the following steps to focus on resistance to land dispossession and environmental injustice. First, it highlights our contribution to dispossession and environmental justice scholarship. Second, it presents the three phases of the *Phulbari Movement*. Finally, we present data to discuss and analyse the *Phulbari Movement*as an anti-dispossession environmental justice movement in neoliberal Bangladesh.

Theory: Dispossession and Environmental Justice

The *Phulbari Movement* offers a rich example of people's resistance to land dispossession and adverse socio-environmental consequences affecting marginalized locations in the global South. Although scholarship on environmentally destructive projects and activities continues to expand, we need more research explicitly connecting land dispossession and environmental justice issues (see Goyes & South, 2016), rather than studying them separately. We use the Phulbari Movement to make a case for studying land dispossession and environmental justice together to grasp the forces driving local politics of resistance in the global South. Below, we discuss a few other vital areas on which scholars must shed critical light to capture such politics of resistance fully.

Although land dispossession has long been a contested development issue, scholarship on it has grown significantly since Harvey (2003) popularized the term "accumulation by dispossession", spurring an enormous volume of research on what is popularly dubbed "global land grabs" (see, i.e., Andreas et al., 2020; Borras et al., 2011; Borras & Franco, 2013; Chuang, 2020; Gellert, 2019; Jay Chen, 2020; Hall et al., 2011; Levien, 2018; Li & Pan, 2021; Nielsen, 2018; Veltmeyer, 2019; Wolford et al., 2013). Harvey's discussion is rather vague and abstract when empirically examined (see Levien & Upadhyay, 2021). Additionally, most scholarship on land dispossession only obliquely mentions environmental justice aspects. One specific example is Levien's (2018) work on land dispossession in India. Levien (with Upadhay, 2021) states that "our study is different, first, in that we examine opposition to land dispossession specifically rather than environmental and other risk factors that may be associated with large capital investments" (2021, p. 6). The Phulbari Movement displays why we need combined theoretical insights of land dispossession and environmental justice to elucidate our understanding of the green potential of many local and grassroots resistances in marginalized areas.

Environmental justice broadly underscores how marginalized groups across several lines are exposed to a higher level of ecological problems (Bullard, 1996; Čapek, 1993; Cutter, 1995; Deb, 2021; Mohai et al., 2009; Sze & London, 2008; White, 2013). Environmental justice scholars initially concentrated on race and class, and their analysis has become intersectional over time, emphasizing higher exposure across several lines, including class, race, gender, nationality, and indigeneity. Globally, environmental justice scholars shed light on the ways in which core, industrialized countries offshore environmentally harmful activities in, or cause harm to the environment of, the peripheral countries (Deb, 2021; Goyes, 2019; Wallerstein, 2004).

We identify two areas of improvement in environmental justice scholarship. The first underscores the ways in which most environmental justice scholars, with few exceptions (e.g., Goyes 2019; Pellow, 2018), largely discount the process of marginalization by land dispossession, even though it is in most cases not isolated from the environmental destruction affecting marginalized groups in the global South. Second, scholars must pay more attention to the instances of environmental injustice in the remote regions of the global South, to offer a more comprehensive account of the issue at stake. Relatedly, to comprehend multiple forms of environmental injustice across the world, David Pellow extends it to many axes of inequalities and sheds light on "the rule of scale", embracing "multiscalar spatial and temporal analyses" (2018, pp. 21–22). It is pertinent to add that scholars from the global South have long demonstrated that environmental struggles and livelihood are closely linked (Baviskar, 1995; Goyes, 2019; Guha, 1989).

Additionally, Deb (2020a, 2021) underscores that slow violence in Bhopal—violence that happens gradually over time and space—is a conceptual example of a multiscalar analysis of environmental justice issues (also see Brisman, 2018). Similarly, other scholars of "environmentalism of the poor" (e.g., Guha & Martínez Alier, 2013) accentuate the ways in which environmental problems affecting minority and the poor of the global South. Furthermore, scholars must heed the specific cases in which marginalized groups in the global South suffer but successfully resist projects to protect their lives, livelihood, and environment.

The *Phulbari Movement*, which we describe in the following sections, exemplifies the importance of studying dispossession and environmental justice together. Such local resistances are successful contestations of the state–corporate's top-down narrative of justice and development. To explain such local and grassroots contestation, we need a framework that underscores a combination of the desire of marginalized people to protect society, land, and the environment. In the following sections, we first describe the movement's three phases and then show how the Phulbari activists remained determined to resolve the problems of dispossession and environmental injustice springing from the largest-ever mining project in Bangladesh.

Three Phases of the *Phulbari Movement*

Phulbari is situated in the Dinajpur district of the Rangpur division of north-western Bangladesh and is one of the most impoverished regions in the country. It is one of the country's most densely populated areas, with 2200 people per square mile. The primary source of livelihood for the people of this region is agriculture: approximately 70 percent of the population is involved in agricultural activities. Due to its vast crop production every year, the area is known as the "bread-basket or "rice capital" of the country (Faruque, 2018b; Hasan, 2020). Nevertheless, more than half of people live below the poverty line. The area is located north of the Himalayan valley and has thus enormous groundwater reserves. The *Phulbari Movement*, born in the region, has captured the public imagination by successfully challenging a destructive neoliberal project in the area.

The Phulbari Movement can be roughly divided into three phases, based on the campaigns in three successive governments from 2005 to 2015: the Khaleda Zia administration (2005–2007); the interim period, backed by the military, during the transition from one elected government to another (2007–2009)[1]; and the Sheikh Hasina administration (2009–2015). The movement faced repression in all three phases, eventually forcing the Hasina administration to abandon the project in 2015.

The first phase lasted until 2007, when the government agreed with the activists to rescind this mining project. The activists regularly worked on several simultaneous campaigns, such as road marches, rallies, assemblies, street protests, and mass petitions, to make people aware of the project's detrimental effects. Receiving support from the National Committee to Protect Oil–Gas-Minerals-Resources-Power and Ports (hereafter the National Committee) was one of the major turning points of the movement. In March 2006, the Committee embarked on a three-day-long march which gathered people from across the country to Phulbari to debunk Asia Energy's win–win narrative (Luthfa, 2019; Muhammad, 2009). In August 2006, after massive publicity of the road march, the National Committee called for a *Gherao* (blockade) campaign in which more than 70,000 people encircled Asia Energy's office (Luthfa, 2011). The protest became violent. The Border Guard of

[1] The interim/caretaker government of Bangladesh was a form of special government led by the non-elected and non-partisan members of the country. The provision of caretaker government was brought out by the 13th amendment of the constitution to rule the country for an interim period of transition from one elected government to another. The prime responsibility of the Caretaker Government is to create a level playing field for each party to ensure a free and fair election within 90–120 days. During its tenure, the CT government could not take any policy decisions. However, as discussed, the military backed caretaker government of 2007 violated most of these rules and remained in power for approximately two years. In 2011, the Hasina-led government abolished the provision of caretaker government through passage of the 15th amendment of the constitution (see Zaman, 2017).

Bangladesh (BGB), a paramilitary force of the Bangladeshi government, shot protestors during the gathering, killing three people on the spot and injuring hundreds of others (Chowdhury, 2019).

The government's brutal squelching further mobilized the crowd. The activists completely blocked the entrance to the Phulbari region from all other parts of the country. They crafted a seven-point demand, the most important of which included (a) banning Asia Energy in Bangladesh and stopping all open-pit mining in the country; (b) compensating the victims of police firing; and (c) building a memorial for the three deceased protesters (Hasan, 2020; Nuremowla, 2016). The then-former opposition leader and current Prime Minister of Bangladesh, Sheikh Hasina, extending her full support to these demands, pledged that her party (the Bangladesh Awami League), if elected, would not carry out any open-pit mines in Bangladesh. At the end of its tenure, the Khaleda Zia administration signed the Phulbari agreement, stating that Asia Energy would leave Bangladesh and that there would be no open-pit mining in the future (Muhammad, 2009).

The second phase of the movement, which lasted only for two years without much activity, started in January 2007, when a controversial military-backed interim government was in power. Despite the 2006 Phulbari agreement, the military-backed government used the shortage of electricity in the country to justify restarting the project. The democratic rights of people under this provisional military arrangement were severely curtailed, and it further worked to instil fear among local activists. In 2007, the Rapid Action Battalion (RAB),[2] created by the Khaleda Zia administration—and notorious for its extrajudicial killings—detained and tortured two frontline Phulbari activists. This interim government also detained the chairpersons of two major political parties in Bangladesh to thwart any effort against the force of the military government. There were no notable protests during the military-backed government.

The movement took a different shape in the third phase during the Sheikh Hasina administration that secured a two-thirds majority in the

[2] The RAB is a paramilitary force under the Armed Police Battalion of Bangladesh Police. It started in 2003 as a force to take rapid action against terrorism and criminality. It consists of members of the Police, Army, Navy, Air Force, Border Guard, and Ansar.

parliament in the 2008 general election. The newly elected government used its legitimacy to advance neoliberal "development" goals in Bangladesh. Despite her earlier promise to the Phulbari activists, the Sheikh Hasina administration resumed the Phulbari coal mine project. The administration linked it to its vision that, by 2030, about 25 percent of the country's energy supply source would come from domestic coal (Hasan, 2020).

The activists also reinvigorated their campaigns. For instance, in October 2010, they arranged a long march, and in February 2011, protesters blockaded a highway to put pressure on the government to rethink the country's development agenda as well as to accept their seven-point demands that the previous administration agreed to implement. However, the government deployed the RAB to suppress the movement. Responding to the threat, in November 2012, the National Committee declared a two-day general strike across Bangladesh, halting trains, blocking roads, and closing businesses and schools across the country. The government, in response, implemented Section 144 of the constitution in Phulbari, banning all assemblies. Despite police attacks, thousands of protesters gathered on the streets to strike. After a relentless decade-long campaign, the Bangladeshi government in 2015 backed away from this project. The *Phulbari Movement* thus shows why neoliberalism in the global South must be studied as a contested terrain in which some forces, such as the ones that threaten the land, livelihood, and the environment of local communities, face more organized pushback than others. The following section further substantiates this claim.

Analysis: Politics of Dispossession and Environmental Suffering in Bangladesh

The proposed coal mining project in the Phulbari region generated a powerful resistance movement that Bangladesh had not seen since its independence in 1971. The government and the corporation marketed the economic and energy potential of this open-pit coal mine project, despite overwhelming evidence that the coal mining industry in Phulbari would have devastated this agricultural, impoverished region. As 70 percent of the people of this region are involved in agriculture-related activities, local communities understood that the coal mine would transform the "rice capital" of the country into a "coal capital". Local communities, including many Indigenous groups, urged the government to withdraw the project. Shantu Larma, an Indigenous national leader, stated:

> We cannot accept this open-pit mining project in this heavily populated *Adibasi* [Indigenous] region. Not only *Adivasis*, I believe, but other communities also will not accept this extremely destructive project. The government must consult with *Adivasis* before constructing any development project in their area. (Faruque, 2018a, p. 79)

However, Asia Energy and the government's key interest was profit (presented as "economic growth potential"). And the powerful actors attempted to employ a false narrative of grassroots favour for the project. Specifically, although Asia Energy reported that the coal mine would affect "household incomes generated from agriculture-related activities" and "landowners, sharecroppers, persons leasing land and agricultural labour who depend on agriculture to survive", the company also declared that about 80 percent of people of local communities think "the coal project is necessary for the development of the country" (Asia Energy, 2006, p. 35).

The way in which the corporation characterized local communities, as described before, contradicts reality. Moreover, three successive administrations attempted to repress the *Phulbari Movement*, exemplifying how the state serves the interest of the powerful, such as corporations, reinforcing multiple forms of inequality (Burawoy, 2017; Carroll & Jarvis, 2015; Evans, 1997; Harvey, 2003; Pellow, 2018; Scott, 1998; Swamy, 2021). The government's crackdown on the 2016 march furthered strengthened the activists' motivation to continue the fight. It had become more apparent to them that the neoliberal Bangladeshi administration, a local activist said, "would do anything in the interest of the rich" and to secure large foreign investments in the country.

The activists henceforth launched a campaign to make people further aware of the adverse consequences of this coal mine. The movement gained further traction when the National Committee, a leftist organization fighting to protect natural resources in Bangladesh, started organizing a national response to this coal mine project. The National Committee then formed the Phulbari Rakhsha (Protection) Committee, which launched a campaign to popularize the idea that land is invaluable to the inhabitants of this region, as it has both material and symbolic value to them. W. Mardi, an Indigenous Santal farmer, raised the concern in the following statement:

> There was no problem ... if that would improve our lifestyle and would benefit the country ... Hundreds of thousands of Bangalees and *Adivasis* [Indigenous] of the Phulbari mining region would have been displaced if the mining had been developed. We would not get a similar life in our new place. Since the open-pit mining project would affect our livelihoods, we strongly protested. (Hasan, 2020, p. 123)

The activists started using new tactics to further organize people around the issue. The 2007 killing of 14 farmers in neighbouring West Bengal, India, who refused to give their land to private corporations (Levien, 2018), might have instilled fear in local communities in Bangladesh. However, it spurred them to improve their tactics and strategy. The activists prepared an "easy-to-follow" pamphlet listing the mine's adverse effects. Such tactics to recruit well-informed activists were useful in generating a mass political movement against harmful neoliberal capital projects in general and the foreign mining corporation Asia Energy in particular.

The company and the government's promise to adequately compensate local communities failed to be credible. Local communities were familiar with many previous instances of dispossession and environmental damage by mining industries. Two, in particular, located in nearby districts, are Barapukuria Underground Mining and Madhyapara Hardrock Mining. Residents were legitimately suspicious that the dispossession and environmental damage caused by large-scale pit mining would never be compensated. Citing two previous cases, a local activist raised his suspicion by saying that "people whose properties were acquired for mining [Barapukuria and Madhyapara] have not received proper compensation as companies promised" (Hasan, 2020, p. 123).

However, due to the administration's commitment to neoliberal ideology, the government time and again showed its complacency with the mining corporation Asia Energy, which, likewise, abused science to misrepresent reality and shift the attention from destructive potential to economic potential. For example, in response to a question about people's apprehension of this destructive project, the CEO of Asia Energy stated: "coal under the ground is worth more than growing rice on the surface".[3] A leader of the National Committee responded to this in the following manner:

> We started working on the issue, contacted responsible ministries, examined a geological survey of the project, and after our investigation, we have not found any related document [from the independent study]. The government had no document related to this project. They did not conduct any study, did not have any assessment report. The government completely relied on the documents made by the company. (Hasan, 2020, p. 127)

Despite the overwhelming evidence of coal as an environmentally destructive fuel, the three succeeding administrations proceeded with Asia Energy's Phulbari coal mine project to encourage foreign capital investment in Bangladesh and ensure their vested interest was realized. Grassroots activities prolonged their fight and expressed their frustration, as reflected in the statement by an Indigenous activist:

> Some advantaged people of the country [policy and decision-makers] ... make plans, lease out natural resources to foreign corporations, and displace and relocate poor people. Some of them already made colossal money ... This project would never benefit people who live in Phulbari ... The majority of people have to lose everything. (Luthfa, 2011, p. 157)

[3] See the BBC report: http://news.bbc.co.uk/2/hi/business/5080386.stm. The same article reported that "the proposed mining area is the size of Manhattan and will affect more than 100 villages—among them communities of the Santal, Munda and Mahali tribes—and the destruction of acres of fertile agricultural land".

The activists remained resolute in continuing the campaign. The *Phulbari Movement* has become the quintessence of the politics of suffering in marginalized locales in neoliberal Bangladesh. It is the largest and longest anti-mining anti-corporate movement in Bangladesh. It is crucial to recall that the-then opposition leader Sheikh Hasina, known as the force of secularism within mainstream politics in Bangladesh, promised in 2006 to rescind the Phulbari project if elected. However, the Hasina administration followed the footsteps of the right-wing BNP government and the military-backed interim government.

The activists realized that their state had abandoned them. Thus, the only way to materialize the movement's demand was by further strengthening their mobilization. They launched a series of campaigns that included a long march in 2010, a highway blockade in 2011, and a nationwide strike in 2012, bolstering their focus on the issue of dispossession and environmental injustice this proposed project would cause. Although the government tried to quell the protesters by imposing the 144 section of the constitution, prohibiting the assembly of four or more people in designated areas, the Phulbari Movement activists never surrendered. Eventually, after a relentless campaign, the government in 2015 decided to back away from the project.

The *Phulbari Movement*, a local farmer named Faizullah states, is a struggle against forces that harm society, land, environment, and livelihood because, as he alludess: displacement will destroy our life. My ancestors cultivated it [land]; then my father inherited part of it ... Now I am using the part I have received from my father; my offsprings will do the same after me, (see Hasan, 2020). An Indigenous leader echoed the same spirit: "I would get ashes and burning heat. It is not only people of the mining region but also the surrounding communities that would be directly affected." The *Phulbari Movement*, as an anti-dispossession environmental justice movement, created powerful local politics in Bangladesh that has become a remarkable symbol of resistance against neoliberal development projects in the global South.

Conclusion

We have analysed the *Phulbari Movement* that forced the Bangladeshi government to abandon a destructive development project as an anti-dispossession environmental justice movement. The movement, in its essence, resisted the project's potential to cause dispossession and environmental injustice. The planned coal mine project, led by British coal mine company Asia Energy, faced organized resistance from 2005 until 2015, spanning three consecutive Bangladeshi administrations. As we have discussed, the communities mobilized to stop the project, which would have adversely affected nearby villages, dispossessed hundreds of thousands, polluted vast water resources, and rendered thousands of acres of agricultural lands infertile. Emphasizing the movement's focus, objectives, and determination, we have brought together both land dispossession and environmental justice issues to illuminate the global South's many specific instances of resistance.

Protests against dispossession and environmental suffering are growing worldwide. Specifically, marginalized people stand up against coal mining projects that do not directly benefit them. Scholarship in this area is also proliferating. We propose that while underscoring such resistance movements is noteworthy, it is equally important to highlight the specific kinds of local and grassroots political opposition to mining in the agricultural region of the global South. The most popular mobilization in the recent history of Bangladesh, the *Phulbari Movement*, illustrates the importance of such specificity in the context of neoliberalism in the global South. It also offers avenues for future research in this area.

Specifically, continuous research on coal mining is necessary to exemplify the false narrative of progress toward sustainability in the global South. The movement, in this regard, shows the green potential of local activism against dominant actors' so-called economic growth narrative. In conclusion, it is imperative to underscore that blaming the global South for hosting coal mining corporations instead of paying attention to the global economic structure facilitating such environmentally disruptive projects in the global South ignores history and context. Finally, to build a just world and to find resolutions in this age of global ecological urgency, we need a shift in power (Brand et al., 2020; Evans, 2008;

Fraser, 2014; Newell, 2021) and to learn from local politics of resistance, such as the Phulbani Movement. To expand our conversation of critical socio-environmental justice, we must not overlook the factors forcing marginalized communities to become the influential actors of resistance to coal mining in the global South.

References

Adnan, S. (2013). Land grabs and primitive accumulation in deltaic Bangladesh: Interactions between neoliberal globalization, state interventions, power relations and peasant resistance. *The Journal of Peasant Studies, 40*(1), 87–128.

Andreas, J., Kale, S. S., Levien, M., & Zhang, Q. F. (2020). Rural land dispossession in China and India. *The Journal of Peasant Studies, 47*(6), 1109–1142.

Energy, A. (2006). *Phulbari coal project: Public consultation and disclosure plan.*

Baviskar, A. (1995). *In the belly of the river: Tribal conflicts over development in the Narmada Valley*. Oxford University Press.

Bedi, H. P. (2015). Right to food, right to mine? Competing human rights claims in Bangladesh. *Geoforum, 59*, 248–257. https://doi.org/10.1016/j.geoforum.2014.08.015

Borras, S. M., Jr., Hall, R., Scoones, I., White, B., & Wolford, W. (2011). Towards a better understanding of global land grabbing: An editorial introduction. *The Journal of Peasant Studies, 38*(2), 209–216.

Borras, S. M., Jr., & Franco, J. C. (2013). Global land grabbing and political reactions 'from below.' *Third World Quarterly, 34*(9), 1723–1747.

Brand, U., Görg, C., & Wissen, M. (2020). Overcoming neoliberal globalization: Social-ecological transformation from a Polanyian perspective and beyond. *Globalizations, 17*(1), 161–176.

Brisman, A. (2018). Representing the "invisible crime" of climate change in an age of post-truth. *Theoretical Criminology, 22*(3), 468–491.

Brulle, R. J., & Pellow, D. N. (2006). Environmental justice: Human health and environmental inequalities. *Annual Review of Public Health, 27*, 103–124.

Bullard, R. D. (1996). Environmental justice: It's more than waste facility siting. *Social Science Quarterly, 77*(3), 493–499.

Burawoy, M. (2017). Social movements in the neoliberal age. In *Southern Resistance in Critical Perspective* (pp. 35–49). Routledge.

Čapek, S. M. (1993). The "environmental justice" frame: A conceptual discussion and an application. *Social Problems, 40*(1), 5–24.

Carroll, T., & Jarvis, D. S. (2015). The new politics of development: Citizens, civil society, and the evolution of neoliberal development policy. *Globalizations, 12*(3), 281–304.

Chowdhury, N. S. (2019). *Paradoxes of the popular: Crowd politics in Bangladesh.* Stanford University Press.

Chowdhury, K. (2018). Indigeneity, alternative development and conservation: Political ecology of forest and land control in Chittagong Hill Tracts, Bangladesh. In L. Rights (Ed.), *Biodiversity conservation and justice* (pp. 127–147). Routledge.

Chuang, J. (2020). *Beneath the China boom: Labor, citizenship, and the making of a rural land market*. University of California Press.

Cutter, S. L. (1995). Race, class and environmental justice. *Progress in Human Geography, 19*(1), 111–122.

Deb, N. (2020a). Law and corporate malfeasance in neoliberal India. *Critical Sociology*, 46(7–8), 1157–1171. https://doi.org/10.1177/0896920520907122

Deb, N. (2020b). Corporate capitalism, environmental damage and the rule of law. In *Routledge international handbook of green criminology* (pp. 367–381).

Deb, N. (2021). Slow violence and the gas Peedit in neoliberal India. *Social Problems*. Online First. https://doi.org/10.1093/socpro/spab058

Evans, P. (1997). The eclipse of the state? Reflections on stateness in an era of globalization. *World Politics, 50*(1), 62–87.

Evans, P. (2008). Is an alternative globalization possible? *Politics & Society, 36*(2), 271–305.

Faruque, M. O. (2018a). Mining and subaltern politics: Political struggle against neoliberal development in Bangladesh. *Asian Journal of Political Science, 26*(1), 65–86.

Faruque, M. O. (2018b). The politics of extractive industry corporate practices: An anatomy of a company-community conflict in Bangladesh. *The Extractive Industries and Society, 5*(1), 177–189.

Feldman, S., & Geisler, C. (2012). Land expropriation and displacement in Bangladesh. *The Journal of Peasant Studies, 39*(3–4), 971–993.

Fraser, N. (2014). Can society be commodities all the way down? Post-Polanyian reflections on capitalist crisis. *Economy and Society, 43*(4), 541–558.

Gellert, P. K. (2019). Neoliberalism and altered state developmentalism in the twenty-first century extractive regime of Indonesia. *Globalizations, 16*(6), 894–918.
Goyes, D. R. (2019). *Southern green criminology: A science to end ecological discrimination.* Emerald.
Goyes, D. R., & South, N. (2016). Land-grabs, bio-piracy and the Inversion of Justice in Colombia. *British Journal of Criminology, 56*(3), 558–577.
Guha, R. (1989). *The unquiet woods: Ecological change and peasant resistance in the Himalaya.* University of California Press.
Guha, R., & Martínez Alier, J. (2013). *Varieties of Environmentalism: Essays North and South.* Taylor & Francis.
Hall, D., Hirsch, P., & Li, T. M. (2011). *Introduction to powers of exclusion: Land dilemmas in Southeast Asia.* University of Hawai'i Press.
Harvey, D. (2003). *The new imperialism.* Oxford University Press.
Hasan, M. M. (2020). Mining conflict, indigenous peoples and environmental justice: The case of Phulbari coal project in Bangladesh (Ph.D. Thesis, York University).
Jay Chen, C.-J. (2020). Peasant protests over land seizures in rural China. *The Journal of Peasant Studies, 47*(6), 1327–1347.
Kirsch, S. (2014). *Mining capitalism.* University of California Press.
Levien, M. (2018). *Dispossession without development: Land grabs in neoliberal India.* Oxford University Press.
Levien, M., & Upadhyay, S. (2021). Toward a political sociology of dispossession: Explaining opposition to capital projects in India. *Politics & Society.* https://doi.org/10.1177/00323292211016587
Li, H., & Pan, L. (2021). Expulsion by pollution: The political economy of land grab for industrial parks in rural China. *Globalizations, 18*(3), 409–421. https://doi.org/10.1080/14747731.2020.1724244
Luthfa, S. (2011). *"Everything changed after the 26th": Repression and resilience against proposed Phulbari coal mine in Bangladesh.* University of Oxford, Department of International Development.
Luthfa, S. (2019). Showcasing environmental justice movements from the South: Comparing the role of media in Bangladesh. *Society and Culture in South Asia, 5*(2), 290–328. https://doi.org/10.1177/2393861719845168
Mohai, P., Pellow, D., & Roberts, J. T. (2009). Environmental justice. *Annual Review of Environment and Resources, 34*, 405–430.
Muhammad, A. (2009). Grameen and microcredit: A tale of corporate success. *Economic and Political Weekly*, 35–42.

Muhammad, A. (2014). Natural resources and energy security: Challenging the 'resource-curse'model in Bangladesh. *Economic and Political Weekly*, 59–67.
Newell, P. (2021). *Power shift: The global political economy of energy transitions.* Cambridge University Press.
Nielsen, K. B. (2018). *Land dispossession and everyday politics in rural Eastern India.* Anthem Press.
Nuremowla, S. (2016). Land, place and resistance to displacement in Phulbari. *South Asia Multidisciplinary Academic Journal*(13). https://doi.org/10.4000/samaj.4113
Pegu, M. (2011). The Phulbari movement: Resisting Neoliberalism in Bangladesh. *Sustainability and Ownership in People Centred Advocacy in South Asia*, 21–37.
Pellow, D. N. (2018). *What is critical environmental justice?* John Wiley & Sons.
Paprocki, K. (2021). *Threatening dystopias: The global politics of climate change adaptation in Bangladesh.* Cornell University Press.
Schlosberg, D., & Carruthers, D. (2010). Indigenous struggles, environmental justice, and community capabilities. *Global Environmental Politics, 10*(4), 12–35.
Scott, J. C. (1998). *Seeing like a state: How certain schemes to improve the human condition have failed*. Yale University Press.
Swamy, R. (2021). *Building back better in India: Development, NGOs, and artisanal fishers after the 2004 Tsunami.* University of Alabama Press.
Sze, J., & London, J. K. (2008). Environmental justice at the crossroads. *Sociology Compass, 2*(4), 1331–1354.
Veltmeyer, H. (2019). Resistance, class struggle and social movements in Latin America: contemporary dynamics. *The Journal of Peasant Studies, 46*(6), 1264–1285.
Wallerstein, I. M. (2004). *World-systems analysis: An introduction.* Duke University Press.
White, R. (2013). *Environmental harm: An eco-justice perspective.* Policy Press.
Wolford, W., Borras, S. M., Jr., Hall, R., Scoones, I., & White, B. (2013). Governing global land deals: The role of the state in the rush for land. *Development and Change, 44*(2), 189–210.
World Bank. 1995. *Bangladesh: From stabilization to growth.* Washington, D.C.
Zaman, K. S. (2017). *The politics and law of democratic transition: Caretaker government in Bangladesh.* Routledge.

8

Latin American Green Criminology and the Limits of Restorative Justice: An Analysis of the Samarco Case

Cristina Rego de Oliveira, Daniela Arantes Prata, and Bruna dos Santos L. da Silva

Introduction

The complexity of environmental disasters and the difficulties of their reparation are widely recognised in the global South. However, few studies focus on the specific characteristics of environmental victimisation in Southern countries and how alternative dispute resolution mechanisms should be structured to properly repair human and non-human harms. Considering this gap in the literature, this chapter examines the

C. R. de Oliveira (✉)
University of São Paulo, São Paulo, Brazil
e-mail: cris.regodeoliveira@gmail.com

D. A. Prata
London School of Economics, London, UK

B. S. L. da Silva
Carleton University, Ottawa, ON, Canada
e-mail: bruna.santos.silva@usp.br

D. R. Goyes (ed.), *Green Crime in the Global South*, Palgrave Studies in Green Criminology, https://doi.org/10.1007/978-3-031-27754-2_8

challenges and limits of applying restorative justice processes in cases of mass environmental victimisation.

To do this, we analyse one of the biggest environmental disasters in Brazilian history: the Samarco case, related to the Fundão Dam collapse, which caused mass destruction throughout the entire Doce River basin and whose victims, after more than six years of the disaster, still struggle to receive adequate redress. Such analysis aims to contribute to the understanding of Southern green criminology in Latin America by paying attention to the particularities of environmental victimisation and the difficulties of offering appropriate redress in Brazil. The case highlights, for instance, (i) the existence of historical colonial exploitation of natural resources and social inequalities intrinsically linked to environmental harm; (ii) institutional fragility, weak regulation, and victims' difficulty in accessing substantial justice; (iii) contexts of economic dependence as limitations to the reparation of people and nature;, and (iv) the challenges of implementing multi-stakeholder models of reparation in asymmetric and litigating contexts.

Green Criminology and Environmental Restorative Justice in Brazil: Addressing Its Challenges and Limits

Restorative justice (RJ) has many definitions, varying according to different geographical contexts and periods, and depending on the theoretical and ideological plural perspectives adopted by scholars to understand the subject (Pali, 2016, p. 50). For instance, it can be understood as a movement, a philosophy, an approach, a process, a new justice paradigm, or a new tool to deal with conflicts. One of the most used RJ concepts was put forward by Tony Marshall, defining it as "a process whereby parties with a stake in a specific offence collectively resolve how to deal with the aftermath of the offence and its implications for the future" (Marshall, 1999, p. 5).

Despite the lack of consensus on the definition of RJ, the majority of scholarship in the field can agree on core values and principles that must be respected to preserve what RJ needs to achieve—such as the encounter between those involved in the conflict, the reparation of material and psychological harm, and the transformation of society (Johnstone & Van Ness, 2007, p. 9). Thus, the restorative approach entails the following elements: (a) participation of the involved parties, families, and community; (b) attention to the needs of the victim and offender; (c) reparation of the harm caused; and (d) sharing responsibilities and obligations amongst offender, victim, families, and community for overcoming causes and consequences of the harmful events (Campos & Oliveira, 2021, p. 149). Braithwaite highlights that the essential principles that RJ processes must respect are: non-domination; empowerment; honouring legally specific limits on sanctions; respectful listening; equal concern for all stakeholders; accountability and appealability; and respect for fundamental human rights (Braithwaite, 2003).

RJ can be applied to different types of conflicts through dialogue-based methodological tools, such as victim–offender mediation (predominant in European countries), conferences, sentence circles, or circular processes in general (peace-making circles). Restorative programmes can intersect with the criminal justice system for adults and juveniles in various ways or functions: it can be an alternative pathway, a complementary response to the wrongdoing, or an independent process applied in social spaces (schools and universities, workplaces, families, community in general) (UNODC, 2020, p. 24). Conceptualised as a participatory model, "any process in which the victim and the offender, and, where appropriate, any other individuals or community members affected by a crime, participate together actively in the resolution of matters arising from the crime, generally with the help of a facilitator" can be restorative (UNODC, 2020, p. 5).

Nils Christie proposed the philosophical background that sustains RJ in the article "Conflict as Property" (1977), being applied first by European, Oceanic (New Zealand and Australia), and North American countries (the USA and Canada). In Latin America, RJ practices have been applied in Chile, Argentina, Colombia, Ecuador, and other countries in different legal fields, social scenarios, and to different audiences, and with or without a legislative framework systematising its application.

In Brazil, the subject has been a relatively novel practice in development by the Judiciary Power since 2005.[1]

From this perspective, we are still learning and understanding how plural experiences are taking place in the Latin American RJ scenario—as well as in other extra-judicial mediation processes—and aiming to improve the guidelines that frame the subject and reduce the risk of its wrongful application. The methodologies (qualitative and quantitative) adopted by RJ protagonists to evaluate experimental practices must be supported by a sufficient grounding in theory, research, and experience from the body of national literature that comprises the RJ canon.

RJ has been applied mostly to interpersonal crimes, promoting a face-to-face encounter to deal with the aftermath of the conflict between the stakeholders involved. However, its application to complex conflicts—such as those related to crimes against the environment—requires a different approach since the people involved, the human and non-human victims, and the reparation are difficult to address. Green criminology perspectives adopted in the global South can be a useful guide to developing an RJ proposal for conflicts that harm nature.

Green Criminology

Green criminology addresses the most diverse criminal conduct that can be considered crime against the environment (Lynch & Long, 2022). It aims to analyse the dynamics of environmental harm, its regulation, criminalisation, and processes of social control (White, 2008). Green criminology also seeks to understand patterns, motives, and explanations of how harm is caused and what its social consequences are (Wolf, 2011, p. 502).

A distinction between traditional criminology and green criminology is that the former connects the deviant behaviour of an individual to the crime committed through a more straightforward relationship of liability

[1] Besides some initial experiences with RJ which were identified in the 1990s in *favelas* in Rio de Janeiro city, the first steps of its formal development started in 2005 when experimental projects were created in juvenile and criminal jurisdictions in different Brazilian cities (Oliveira, 2021a, 2021b).

and clearer dimensions of harm (as, for instance, in the case of homicide or financial fraud). In instances of environmental harm, its causes cannot always be easily identified, and its impact may only be felt after a certain period of time. Moreover, Earth is not an infinite resourceful expanding economy, and environmental harm may take years to be restored and, in some cases, it may never return to its previous state (Brisman & South, 2017).

Criminological studies of environmental crime from the global South have been overlooked by the mainstream literature on the field for decades, and important characteristics of green criminology in Latin America (such as the legacy of colonisation and the value of the epistemological contributions of the marginalised) have been undervalued (Goyes, 2021). Considering such a scenario and the goal of this chapter, this section will focus on the characteristics of environmental victimisation according to green criminology and understandings of environmental disaster management and reparation processes in Brazil.

Green Criminology, Environmental Victimisation, and Environmental Crime in Latin America

The framework for studying environmental victimisation proposed by green criminologists helps us to understand the impacts of environmental crimes on their victims and the difficulties of their reparation. Hall (2013, pp. 34–36, 127), when proposing a "green victimology", highlighted the complexity of environmental harm, whose impacts: (i) usually go way beyond material damages and cannot be simply repaired by financial compensation; (ii) may cause systemic and long-term social and cultural harm; (iii) victimise the ecosystem and a heterogeneous range of individuals; and (iv) challenge the justice system's ability to provide satisfactory legal responses to damages and crimes.

Similarly, White (2013, pp. 1–6) highlighted the complex and multifaceted nature of environmental harm, which gives rise to issues of recognition, acknowledgement, participation, reparation, and compensation, as well as inequalities among affected communities. An interdisciplinary approach to studying environmental harm is essential, and working with integrated perspectives on different possible interventions may be more effective than focusing on one solution (Gibbs et al., 2010, p. 139).

In Latin America, the study of environmental harm must take into consideration the historical and growing foreign exploitation of natural resources in the area, which include mining activities, deforestation, and conflicts over natives' land for agricultural purposes, which are particularities of the colonial nature of the transnational accumulation of natural goods (Ramos, 2010). Mining and deforestation disturb ecosystems, making them more prone to environmental disasters, whether of human or natural causes (Farber, 2011).

Carvalho and Damacena (2013) emphasise the need to anticipate hazards and risks in disaster management, either when a disaster is natural (and is apparently unrelated to any control but potentially subject to public oversight) or man-made (there being usually a system to exercise certain control over risks). In this sense, risk communication would help rationalise environmental uncertainties in the face of complex events that can be caused by different reasons and have drastic consequences (Farber, 2012).

For Carvalho and Damacena (2013), disasters in Latin America are usually marked by: (i) their multidisciplinary context, (ii) the importance of risk management, and (iii) the scenarios of regulatory failure.

Multidisciplinarity responds to the ecocomplexity of disasters since it affects the same environment on different levels (fauna and flora), which lawmakers alone cannot capture through environmental law and litigation. The identification of risks, the degree of potential impacts, and proposals for reconstruction should be guided by work teams from different areas, such as geology, climatology, meteorology, engineering, architecture, economics, social sciences, and law (Carvalho & Damacena, 2013). Just as understanding the dynamics of environmental harm needs interdisciplinary work, responses to environmental victimisation require going beyond mere legal understanding to include a truly interdisciplinary perspective (Gibbs et al., 2010; Hall, 2013; Orlando & Bergin, 2017; White, 2014).

In turn, risk management is part of the disaster cycle (Farber, 2012): a disastrous event is followed by emergency responses, mitigation, compensation, victim assistance, and reconstruction; and these last phases, while concerned with the restoration of the environment to its previous status, also face risks of a new or an ongoing disaster, creating a constant management cycle (Farber, 2012).

Disaster risks are not something entirely beyond human control. With planning, "the risk of flooding can be reduced, nuclear reactors and offshore oil rigs can be made safer, and climate change can be limited" (Farber, 2012, p.6). In this sense, for Verchick (2010, p. 54), disasters are caused or increased by a regulatory failure to protect the environment—as natural infrastructures would usually act in favour of the protection of life, as for example occurs with the vegetation of slopes with long roots, grasses, and weeds that sustain the land and prevent landslides.

Despite the existence of progressive legal frameworks attempting to protect the environment in countries like Brazil (Prata & Garrido, 2022), weak regulatory institutions and poor enforcement are central to the recurrence of environmental crimes and disasters in Latin America (Saad-Diniz, 2019, pp. 46, 52). In the Brazilian corporate criminal context, poor law enforcement, institutional fragility, and weak regulatory capacity lead to non-collaborative responses from corporations, and thus to inadequate prevention and deterrence capacities (Saad-Diniz, 2019, pp. 46, 52). Weak regulatory institutions and poor enforcement (which indeed are within the corporate criminal liability framework in Brazil), therefore, should be one of the considerations to be attentive to when studying Southern green criminology (Goyes, 2019). An example of weak regulations and accountability for environmental exploitation is the Amazon Cooperation Treaty, in which Brazil, Bolivia, Colombia, and many other Latin American countries agreed on promoting sustainable development in the region, including the quality of life of the local population and the defence of natural heritage. Despite its progressiveness, the agreement does not mention the responsibility of foreign companies that carry out mineral, oil, and gas exploration in the region, as well as those which cause deforestation for monoculture production (Nunes, 2016).

When it comes to disaster redress, Farber (2012) described three means for victim compensation: (i) private insurance; (ii) private lawsuits; and (iii) public lawsuits. The first one may be unfeasible due to possible contractual hurdles and the high cost of the policy, and it depends on the predictability of the disaster risk. Nonetheless, it is possible to find in Brazil several insurance companies that offer coverage for damage caused by floods. The two other compensation methods are based on the possibility of bringing lawsuits against environmental offenders, aiming to redress both the disaster victims and the environmental harm caused by the action, omission, or negligence of private and/or public entities.

In Brazil, civil liability proceedings are the common route for environmental victims to seek redress. However, considering judicial delays and court backlogs, as well as the difficulty of proving the extent of harm and the causality between an offender's action, the environmental harm, and the losses of the victim, extra-judicial mechanisms have become central to individual compensation. Such a scenario is clear in the Samarco case (see section "The Samarco case" below) and highlights the difficulties of redressing victims of environmental crimes—as Hall (2013) and White (2013) described in other cases. This scenario is not exclusive to Latin American contexts, but it is essential to highlight that victims in global South countries—such as Brazil—may have limited access to substantial justice due to judicial delays, court backlog, excessive possibilities of appeal, and lack of appropriate enforcement (Prata & Garrido, 2022).

Finally, the environmental reconstruction phase within the disaster risk management cycle aims to restore the environment to its previous status (before the harmful events took place) and provide a new possibility for the development of the human and ecosystemic activities which used to take place in the affected area (Farber, 2012). On this topic, Brazilian environmental law applies the principle of full reparation of the environment, which imposes on the offender the obligation to restore the site to the same state as it was prior to the harm and, if it is not possible to do so, the offender should make ecological compensation, improving the environment in another equivalent area, preferably at a site which has influence on the original place of harm (Milaré, 2015).

The idea that nature is capable of restoring its ecosystems after several years is not fully accepted in green criminology, especially in Latin America, where environmental degradation increases progressively, and in some scenarios harm is irreversible. Green criminology acknowledges ecosystems and their components as living entities, and includes addressing a diverse array of non-human crimes that affect human and non-human animals alike. Humans make themselves, other animals, and ecosystems victims of their "development". Green criminologists expand the category of victim to also encompass ecosystems and non-human species, understanding that they are also living beings that can be harmed and victimised, and that must receive greater attention in discussions of prevention and restoration.

Brazilian Restorativism and the Absence of Environmental Perspectives

The Brazilian RJ movement is characterised by three hegemonic perspectives related to: (i) the definition of the RJ protagonist actor, (ii) the typology of practices adopted, and (iii) the foreign narratives that inform the phenomenon, appropriated from the global North (Oliveira, 2021a).

Brazilian RJ is led by judiciary power, which finances, implements, and shapes most practices. It has adopted and spread a typology of "peace-making circles" in RJ, a methodology introduced in southern Brazilian states by Kay Pranis, an American practitioner, and implemented to deal with all types of interpersonal conflicts. Having few prominent national pieces of literature on the topic, it is possible to notice our reliance on the knowledge produced by Canada, the USA, and Europe not only regarding the methodological approach adopted (the peace-making circles) but also from the theoretical perspective, having authors such as Howard Zehr (2008), Paul MCold (2001), and Marshall Rosenberg (2006) at the centre of the RJ framework in Brazil. The scenario is problematic, given that "we cannot copy foreign models naively and unselfconsciously, especially from countries whose judicial tradition is different from ours, as are common law countries" (Campos & Oliveira, 2021, p. 149).

Despite the proliferation of the RJ movement around the country, its practices are still applied only to less serious crimes,[2] in which victims, offenders, and damages caused by crime are easily identifiable. Brazilian RJ is not focused on the possibilities of using its practices to deal with the aftermaths of complex crimes—such as those committed against the environment, and which result in mass victimisation.

RJ also faces challenges in this complex environmental field. To promote a systematic transformation that engages civil society, it is necessary to define core elements (values and principles) that sustain a *participatory* decision-making practice, protecting the interests of non-human victims, motivating corporate offenders to take up responsibility and enter into dialogue with harmed communities, and organising appropriate panels for public discussion to make different types of restoration possible. In this sense, strategies need to be settled to define which restorative practice can be applied to specific conflicts, who will participate in the process—how to determine victims, offenders (individuals and corporations), and communities affected by the conduct, and how to represent nature and future generations—and what (types of) harm was done and what are the possibilities for its reparation.

Restorative practice aims to promote voluntary participation and free dialogue between stakeholders involved in the conflict; and in corporate environmental crimes, a key element to a successful practice is assuring the engagement of the interested parties and reducing power imbalances. To do so, it is necessary to determine the legitimate victims/offenders or their representatives, how to empower them during the process, and how to represent nature legitimately and effectively during negotiations. Such concerns align with restorative justice goals and principles, highlighting the role of informed participation and the need for non-domination and empowerment of victims (Braithwaite, 2003).

According to Hamilton (Forthcoming, p. 4), in corporate crimes "the immediate and long-term success of the restorative justice process will

[2] As in Brazil there is no legal framework to underpin RJ processes for adults: some experimental projects were created and are currently being applied to less serious crimes (with two years of deprivation of liberty as maximum penalty); and some experimental projects deal with domestic violence conflicts. The only RJ legal framework in the country concerns the juvenile justice system (Law 12.594/2012).

depend on the knowledge and commitment of these representatives, along with their ability to influence cultural change within the organisation". This also means that individuals representing corporations in restorative processes must be representative enough to show the corporation's genuine interest in restoring harm and its willingness and drive to change. It is thus necessary to further develop RJ theories and their practical discussions to a second level of complexity to promote reflections related to the feasibility of RJ programmes to deal with *serious and complex cases* involving diffuse victimisation and non-measurable damages, urged by the current neoliberal context that is devastating nature and threatening sustainable life for all living creatures, human and non-human.

This neoliberal scenario is worsened by the current Brazilian political agenda, which tends to dismantle national environmental policies by reducing budgets, closing institutional bodies, promoting deregulations, denying high deforestation rates, and refusing to protect Indigenous communities, among others. Despite being a Southern country rich in biodiversity, Brazil has had international headlines due to the genocide of Indigenous communities (Carvalho et al., 2020), the inefficiency of the governmental agencies to curb the Amazon's and the Pantanal's deforestation, the harm caused to marine environments by repeated oil spills, and the tailings dam collapses in the past decade.

Facing a "regulatory rollback"[3] (Saad-Diniz & Gianecchini, 2021, p. 6), the legal frameworks developed to protect the environment in Brazil are insufficient to prevent wrongdoing and promote accountability. In other words, environmental tragedies happen systematically in the country without sufficient legal responses (Saad-Diniz & Gianecchini, 2021) or restorative initiatives. When the legal system cannot reasonably address environmental disputes, alternative conflict resolution mechanisms offer an opportunity to better redress societies impacted by these harms and protect nature. In other words, when criminal, civil, or administrative justice systems are insufficient to deal with the conflict

[3] According to the authors "the rollback of regulations responsible for protecting the environment ... is directly related to the intensification of exploitation of people (labor), land, and the biosphere, which leads to the intensification of environmental devastation" (Saad-Diniz & Gianecchini, 2021, pp. 6–7).

and its consequences, the RJ approach can be an innovative tool to respond to environmental harms, leading victims, offenders, and local communities to a new ethics of care, able to reframe their relationships with nature (and with others) and prevent future damage (Braithwaite et al., 2019, p. 10).

There has been no concrete attempt to apply RJ in the Samarco case, and its reparation process has been far from achieving any RJ goals. Two reasons to analyse the Samarco case using RJ lenses are, however, (1) the extra-judicial dynamics that took place in the aftermath of the disaster and what they have to say for the application of RJ in Latin America; and (2) the recurrent use of extra-judicial mediation in other environmental disasters in Brazil and what the learnt lessons can contribute for the Samarco case.

Latin American Environmental Harm and the Samarco Case

In Latin America, as in other regions of the world, environmental problems arise from several triggering factors. To narrow our proposed analysis and understand the Latin American context in its social and historical aspects, this section will focus on the impact of the state and private corporations, including foreign ones, on the extraction of natural resources in Latin America and the global economic position of the area as a local producer of commodities to export.

Concerning the latter, the current influence of foreign capital in Latin America is closely related to the processes of colonisation (Goyes, 2019). Such processes were initially characterised by the strong influence of Europe, but more recently by the United States, especially since the 1950s and 1960s, in which liberal and developmental thinking transformed industrialisation in Latin America (a situation captured by the Theory of Latin American Dependence [Quijano, 2005]). Already in the 1990s, the commercial deficits of the region were financed with capital inflows driven by internal and external debts, and the industry scenario was influenced by a growing number of foreign extractive companies acting in the area (Ramos, 2010).

The context of economic dependence throughout history also relates to the occurrence of environmental crimes in Latin America. A clear example of how foreign trade affects Latin America's climatic conditions is the Amazon Forest (Vera & Vargas, 2021). The expansion of agricultural monoculture, the construction of roads, and the increasing agricultural and mining activities in the area lead to the destruction of ways of life, forcing people to witness the degradation of their forests or to accept impositions that, in many cases, compromise their integrity and modify their cultural bases. One of these impositions is forced displacement of Indigenous communities, a process of Indigenous deterritorialisation that affects the cultural aspects which determine these social groups.

A further issue to be attentive to in the Latin America extractive industry is the potential context of communities' dependence on companies. According to Saad-Diniz (2019), social relationships and the local economy can be so dependent on companies that community victimisation discussions may be oblique. This situation can increase the vulnerability of victims (Mazzucato, 2018, pp. 58–59), influence power dynamics in legal and extra-judicial proceedings, and put into question the collateral effects of certain sanctions—considering, for instance, that sanctions such as incapacitation (shutting down companies) may do more harm than good to a local community that depends on such activity.

In the Latin American context, the exploitation of the natural resources of a given region, often directed by foreign companies, defines its local economy, and generations of families work with extractive activities to survive. Green criminology in Latin America needs to consider the scenario of coloniality of power (Quijano, 2005), in which the foreigner (or the one that possesses economic power) creates community dependence on a certain activity brought or created by the coloniser's needs, and in which exploitation of the environment becomes a side effect, distributing positions of power and discrimination in society.

The Samarco Case

The Samarco case relates to one of the biggest environmental disasters in Brazilian history: the Fundão dam which collapsed on 5 November 2015, in the rural area of Mariana, a mining city in the State of Minas Gerais (MG). The collapse caused environmental destruction for over 600 km down the Doce River and severely impacted thousands of lives in the affected areas. The dam was owned and operated by Samarco Mineração S.A., a joint venture controlled by two multinational mining giants: the Brazilian Vale S/A and the Anglo-Australian BHP Billiton. Many different accounts of the severity of the impacts and related litigation have already been made, as well as of the mediation and grievance mechanisms implemented in the case.[4]

Over 40 million cubic metres of mining tailings travelled down the Doce River, affecting 46 municipalities, two states, and thousands of people, including several traditional communities. It was a human *and* an environmental tragedy. The collapse resulted in 19 fatalities, among them Samarco workers and residents of Bento Rodrigues, a nearby community overthrown by the tailings, as well as a miscarriage and physical injury to several people. It also led to socio-environmental and socio-economic harms not entirely possible to measure nor repair. Environmental harms included the contamination of water resources and soil, the devastation of ecosystems, the destruction of habitats, and high animal mortality along the river and its basin.

Environmental harms also provoked detrimental effects on productive activities such as agriculture, livestock grassing, fishing, mining, trade, services, and tourism, leading to the impairment of the regional economy and people's way of living in several municipalities, including a rise of unemployment and a decrease of tax revenue. The collapse also destroyed public and private infrastructure in rural and urban areas, and damaged historical and cultural heritage. Socially, the disaster created long-term damages to physical and psychological health, safety, culture, housing, and the social organisation of the affected communities. Traditional

---

[4] The case description in this chapter is based on Prata (2019, 2020) and references therein contained.

groups were also severely impacted, as their historical, social, religious, and cultural relationships to their land and river were profoundly harmed by the disaster, leading to a gradual loss of cultural identity and their traditional lifestyle. Almost seven years after the collapse, in 2022, many victims were still unsatisfied with the resolutions, and the environment had still not been fully restored.

Green Criminology, Restorative Justice Processes and the Samarco Case: Lessons from the Field

The Samarco case presents clear aspects of complex environmental victimisation, such as (i) the variety of types and degrees of different harms to different victims; (i) the difficulty of measuring damages; (iii) the heterogeneous character of the victims; (iv) the need for an interdisciplinary approach; (v) the occurrence of human rights violations; and (vi) the specific victimisation of traditional communities (Prata, 2020).

There are, however, other characteristics of victimisation in the Samarco case that deserve further attention and that, considering the specific Latin American context, can help inform the development of Southern green criminology. The first relates to the victimisation of traditional communities and how particular it can be in the Brazilian extractive context, especially in light of historical human rights violations and the existence of prior environmental conflicts between communities and offenders, the states or third parties.

One example is the Krenak Indigenous community, one of the populations most affected by the disaster. The Krenak people had historically occupied the Doce River basin, and had been victims of several human rights violations throughout history, including recent events in the twentieth century—having been repressed and incarcerated by the military dictatorship in Brazil (sent to the "Reformatório Krenak") and after that arbitrarily displaced (sent to the "Fazenda Guarani") (Governo do Estado de Minas Gerais, 2017). In recent decades, the Krenak community had already been in several disputes over their territory and environmental degradation against Vale—one of the mining companies that owns Samarco. Scenarios like this have several implications for traditional communities in the aftermath of environmental disasters, both in relation to their impacts and the reparation processes.

First, the harm caused to these communities must not only be seen with great attention to all the cultural and traditional impacts they imply, but also in light of historical violations and environmental injustice, adding a further layer of victimisation and complexity to their pain and suffering. Second, the pre-existent context of historical conflict and environmental disputes must be carefully considered in the reparation process, considering that there may already exist trust issues between parties and less interest in restorative circles and negotiated processes. The Krenak community, for instance, for a long time refused to negotiate their reparation with the Fundação Renova, preferring to negotiate with Vale, the company with whom they already had their disputes and ongoing discussions.

Another relevant particularity of victimisation in Latin America that is present in the Samarco case is the context of community dependence on corporations. Before the disaster, the economy of Mariana was very much dependent on mining activities: Samarco was the biggest employer in the city, and approximately 80% of Mariana's income tax came from the mining sector and 54% from Samarco (Grupo da Força-Tarefa de Minas Gerais, 2016, p. 246; Prata, 2019, p. 256). Such a scenario impacted some of the victims' initial response to the disaster, as well as the response from the local community in Mariana, who expressed the need for Samarco's operation to return, despite also demanding justice—with Samarco's "social license to operate" being highly influenced by such a dependence relationship (Mazzola, 2018, pp. 175–177).

This context is also important when analysing reparation processes and extra-judicial negotiations in the case—and in Latin American environmental victimisation scenarios in general—given that the dependency relationship can help create: (i) scenarios of asymmetry and acquiescence between victims and the offender; (ii) conflicts between victims (with regard to their attitude towards the offender's return to activities); and (iii) power imbalance even between government and company, given that municipalities and states may be dependent on returns from the offending company (Prata, 2019, p. 258).

As an initial response to the disaster, legal authorities and offending companies in the case attempted to implement a distinctive form of reparation process to address human and non-human harms caused by the collapse. Six months after the event, they created a (supposedly) independent foundation—the "Fundação Renova"—to repair all harms caused by the collapse, including to the environment, individuals, and public infrastructure. The Foundation was created through the signing of a "Transaction and Conduct Adjustment Term" (TTAC), an agreement between the companies (Samarco and its shareholders, Vale and BHP) and the Brazilian government (the federal government, the state governments of Minas Gerais and Espírito Santo and other government agencies). TTAC also created 42 programmes, divided between socio-environmental and socio-economic projects to handle specific issues and workstreams.

To monitor and review the work of the Foundation, the parties also created the Interfederative Committee (CIF), a public entity formed by members of the government, experts and representatives of municipalities, and other legal entities. CIF were to have regular meetings, debate issues arising in the programmes, and decide on disputes between parties.

The Public Prosecutors Office, victims' movements, and academic institutions heavily criticised TTAC due to the lack of participation of prosecutors in its negotiation, the complete absence of victims' voices, and the proposed cap on reparation values (at the time, valued at BRL20 billion). Several criticisms were also made towards Renova's grievance mechanisms, including the lack of previous, free, and informed consultation of Indigenous peoples; the lack of participation of victims in Renova's and CIF's decision-making processes and in the reparation process in general; the lack of information; the asymmetry of power in extra-judicial negotiations; the inadequacy, slowness, and delay of several programmes; the lack of recognition of several victims; the difficulties of proving damages such as past income and material losses (which often lead to offers of low compensation values in Renova's financial compensation programme); and Renova's non-compliance with CIF's resolutions and directives (Prata, 2019).

Victims' demands for accurate information, participation, and the right to technical assistance were among the most recurrent and emphasised ones. Here, assuring the victims' right to be assisted by technical organisations—so debated in the Samarco case—would empower them. However, the process of hiring such organisations and allowing them to act faced many hurdles in the case, also putting into question its possibility of success.

Another critical aspect of reparation is the need for genuine interest on the part of corporations in restoring communities—which is fundamental in RJ processes. This, unfortunately, was not the scenario in the Samarco case: victims felt that the Foundation was not as independent as it claimed to be, that they were being offered insufficient amounts of compensation, and that the companies' litigating behaviour was far from indicating a genuine interest (e.g. legally questioning CIF's decisions or denying the causation of victims' harms in individual legal proceedings).

The problematic scenario at the Renova Foundation led to further litigation between public prosecutors and the companies, and thus to the long-term negotiation of new agreements, which aimed to improve the governance of Renova's reparation by (i) hiring experts, audit companies, and organisations to provide damage assessments and technical entities to assist victims in the extra-judicial negotiations (agreements were signed in January and November 2017); and (ii) improving victim participation in the reparation process and including this in renegotiating Renova's programmes along with the assistance of independent experts (the "GTAC agreement"). This agreement aimed to create a new governance system which included the creation of at least 19 local commissions through the Doce River basin, which would discuss and negotiate with the Foundation their specific local issues in particular programmes, and propose solutions that, in turn, would be sent to Regional Cameras to be mediated by legal entities (e.g. public prosecutors) and the companies, and then implemented. The local commissions would be created with the help of and receive support from technical assistance organisations, which would also be chosen and hired locally in a participatory process supervised by one of the consulting organisations hired in 2017.

The role of technical organisations would be essential for access to justice in extra-judicial settings: to promote the informed and effective participation of victims, empower them, and reduce the imbalance of power in extra-judicial negotiations (Homa, 2022a). As of September 2022, the process of hiring and fully implementing technical assistance to victims, however, is still far from being a reality, four years after GTAC was signed.

There were different reasons for the delay in this implementation. One reason is the complex management of the hiring process and also the natural time-consuming aspect of the democratic decision-making required for it (Fundo Brasil de Direitos Humanos, n.d.). There were also major disagreements between companies and the technical organisations when it came to the terms, prices, and conditions of the work to be conducted, with offending companies rejecting offers and their hiring process having to be brought to court for the judge to decide. And this illustrates another difficulty in ensuring victims' rights in alternative forms of resolution, especially in countries with a strong litigating culture such as Brazil: (i) the easy and constant use of litigation (and its expected delay) as a negotiation tool; and, once again, (ii) the seeming absence of genuine corporate interest in full reparation.

The parties increasingly used litigation during extra-judicial negotiations. This "judicialisation" of extra-judicial disputes was part of a shift in the case form of resolution in late 2019 and early 2020, when the court started to play a major role in deciding controversies between parties. Throughout 2019, in light of several complaints from the government and public prosecutors about Renova's delays and non-compliance with agreements and CIF decisions, several of Renova's actions, plans, and deadlines were recurrently brought to court. In January 2020, the court evaluated that the agreements were not being adequately complied with and decided to implement a new processual system to the case, creating 11 "Priority Axes Proceedings", each focusing on relevant and controversial topics (such as the hiring of technical assistance organisations), and to which parties should submit their arguments and claims for the judge to decide (Observatório Nacional, 2021).

There is a relationship between judicial and extra-judicial proceedings in the Samarco case: claims are brought to court, and the civil legal response usually ends up becoming extra-judicial and relying on Renova, both in the collective and individual spheres of redress, in order to avoid judicial delays and court bureaucracy. However, the delays in programme implementations by Renova and non-compliance with its agreements, as well as disagreements between parties in the execution of certain programmes, have led to the "judicialisation", again, of the extra-judicial discussions, emptying the reparation governance system first established between the CIF, the Foundation, and legal authorities, and which was supposed to include victims' voices through local commissions.

Such a system was basically abandoned after a court's decision in July 2020, which created an alternative route for victim compensation: the Simplified Indemnification System. On this occasion, the court recognised the failure of Renova's governance system and decided to implement a faster route to individual compensation. For this, the court decided on a new "matrix of damages" and category-based compensation amounts without victim participation and disregarding the reparation governance system, the work of technical assistance organisations, and other legal authorities such as the Public Prosecutors Office (Garrido et al., 2022).

Despite being challenged by legal authorities and questioned by social movements and scholars due to its top-down approach, disregard for victim participation, disregard for the whole prior reparation governance and participatory efforts, and imposition of rights-waiving conditions (see Garrido et al., 2022), the Simplified Indemnification System was put in place and implemented in several locations.

At the same time, the main legal institutions in Brazil—including the National Justice Council—recognised the several failures in the reparation process of the case, and started negotiating with the companies a new and final solution to the disaster, with negotiations supervised by the Council and voices of communities being heard through public hearings (the common public participatory instrument in Brazil). However, and again demonstrating the difficulty of coming to terms with the offending companies, this renegotiation—called the "Repactuation"—failed to provide any results after 14 months, leading to the abandonment of the discussions and the reopening of public litigation in September 2022 (Homa, 2022b; Ministério Público Federal et al., 2022). In a recent joint press release, public authorities justified the situation, claiming that "the attitude of the companies has so far shown a lack of commitment to social and environmental responsibility practices" (Ministério Público Federal et al., 2022).

All these revolving scenarios illustrate further issues for participatory methods of conflict resolution such as RJ. First, the complexity of the identification of multi-stakeholder agreements may imply in delays, conflicts, and victim dissatisfaction and, ultimately, in the rejection of negotiations to seek rights assurance in courts. Second, there seems to be a dependency of parties on the use of legal proceedings if one of them is not genuinely trying to engage in a meaningful and restorative conversation—which can be problematic in countries with issues of access to substantial justice and an intense court backlog, such as in Brazil and other Latin American countries. Third, if an offender does not comply with extra-judicial agreements, parties need to seek enforcement in court anyway; which brings them back to the need to face costs, delays, and potential litigation.

The Reparation of the Environment: A Neglected Victim?

Legal and socio-legal analyses of the Samarco case emphasised the need of repairing individuals (with the environment connected to them) rather than the environment itself. The rehabilitation of the environment is, of course, part of the collective claims, negotiations, litigations, and reparation programmes, with specific reports and attention to technical questions. However, what seems to be missing is a further focus on the rights of nature and its participatory representation in public debates (going beyond its representation by public prosecutors in courts and negotiation rounds).

Conflicts between corporations and communities in the global South are based on systemic injustices, extreme power imbalances, and high victim vulnerability. Dealing with victimisation that has collective and community aspects is a major challenge faced by restorative practices and, in socio-environmental cases, it is needed to address human and non-human dimensions of harm (Aertsen, 2018). Despite the difficulty that diffuse victimisation entails, another issue concerned with ecological and interspecies justice emerges in these cases. Considering that a "diversity of beings and entities" can be harmed, ideally an anthropocentric perspective has to be replaced with (or at least accompanied by) a more ecocentric philosophy[5] (Varona, 2019). However, when it comes to RJ, "victimology itself as a discipline has been constructed on anthropocentric, personalised and individualised notions such as experience, risk, impact, needs, rights and narratives" (Varona, 2019, p. 676).

In fact, since traditional restorative practice is mostly applied to non-complex cases, RJ scholars and practitioners are adopting the same anthropocentric route. Criminological perspectives that defend RJ application to environmental cases include a variety of victims in the practices: in the case of non-human entities harmed, mechanisms such as surrogate victims or "human guardians as representatives" of nature (Hamilton, 2021, p. 226) are being tested as a means of representation. In Latin American countries where communities are vulnerable and economically dependent on big corporations—or indeed are being explored—the delimitation of who should be the legitimate representative of nature is a relevant issue: from a colonial perspective, the reparation of nature can be surpassed by financial or even political interests guided by personal benefits, and "nature empowerment" is as equally important as the empowerment of individuals.

Conclusion

Considering the disparities between practices applied in different Latin American countries, it is essential that practitioners and scholars dialogue over the existing knowledge already produced in the global South to define what RJ should be developed to deal with local environmental issues, which must take into account the historical extractive context of

[5] From a more ecojustice perspective, environmental rights come to the fore from the notion of ecological citizenship and the conceptualisation of species justice where other non-humans can be conceived as victims (Varona, 2019, p. 675).

exploitation and environmental injustice in Latin America and related structural problems such racism, gender, and social inequalities (Goyes, 2019).

From this perspective, a few lessons can be inferred after reflecting on RJ in Latin America, focused on Brazilian viewpoints through the analysis of the Samarco case—which sheds a light on (i) the particular characteristics of environmental crime and environmental victimisation in Brazil, (ii) the challenges of implementing multi-stakeholder models of reparation, and (iii) how the interplay between legal and extra-judicial responses plus the difficulty of accessing material justice in courts may negatively impact extra-judicial resolutions of a case.

First, RJ has not been used to deal with serious and complex cases involving diffuse victimisation and non-measurable harms, especially socio-environmental conflicts. That is partially because it is still necessary to (re)arrange RJ conceptions, re(think) who are the legitimate stakeholders able to participate in the dialogue, observe what should be the dimensions of damage, and, importantly, create innovative possibilities of reparation and victim empowerment. Second, the limitation of victims (human and non-human), participants (or represented), and their lack of voice during the decision-making process are sustained by a range of cultural and historical colonised conditions—such as community dependence, power imbalances between the stakeholders, absence of free informed victims' rights, and overexploitation of environmental resources—that seem to legitimate the corporate indifference to effectively repair the environment and support the empowerment of communities.

References

Aertsen, I. (2018). Restorative justice for victims of corporate violence. In G. Forti, C. Mazzucato, A. Visconti, & G. Stefania (Eds.), *Victims and corporations: Legal challenges and empirical findings* (pp. 235–258). Wolters Kluwer Italia.

Braithwaite, J., et al. (2003). Principles of restorative justice. In A. Von Hirsch (Ed.), *Restorative justice and criminal justice: Competing or reconcilable paradigms* (pp. 1–20). Hart Publishing.

Braithwaite, J., Forsyth, M., & Cleland, D. (2019). Restorative environmental justice: An introduction. In B. Pali & E. Biffi (Eds.), *Environmental justice restoring the future: Towards a restorative environmental justice praxis* (pp. 8–12). European Forum for Restorative Justice.

Brisman, A., & South, N. (2017). *Green criminology* (6th ed.). The Oxford Handbook of Criminology.

Campos, C. H., & Oliveira, C. R. (2021). Judicial restorative justice and domestic violence in Brazil: Setting problems and challenges from the field. *International Journal for Crime, Justice and Social Democracy, 10*(4), 146–157. https://doi.org/10.5204/ijcjsd.2015

Carvalho, D., & Damacena, F. (2013). *Direito dos Desastres no Brasil.* Livraria do Advogado.

Carvalho, S., Goyes, D. R., & Vegh Weis, V. (2020). Politics and indigenous victimization: The case of Brazil. *The British Journal of Criminology, 61*(1), 251–271. https://doi.org/10.1093/bjc/azaa060

Christie, N. (1977). Conflict as property. *The British Journal of Criminology, 17*(1), 1–15.

Farber, D. (2011). Navigating the intersection of environmental law and disaster law. *BYU Law Review, 6*, 1783–1820.

Farber, D. (2012). Disaster law and emerging issues in Brazil. *Revista de Estudos Constitucionais, Hermenêutica e Teoria do Direito, 4*(1), 2–15.

Fundo Brasil de Direitos Humanos. (n.d). *Resumo Executivo—Programa Rio Doce. Resultado do processo de Escolha de Assessorias Técnicas aos Atingidos pelo Rompimento da Barragem de Fundão na Bacia do Rio Doce, nos termos do contrato no. 4500176574.* http://www.mpf.mp.br/grandes-casos/caso-samarco/documentos/fundo-brasil/memorial-das-escolhas-de-assessoria-tecnica..pdf

Garrido, D. B., Prata, D. A., & Abreu, C. M. A. (2022). New route for redress in the Samarco case? An overview of the Simplified Indemnification System's (un)lawfulness. *Business and Human Rights Journal, 7*(3), 475–480. Special issue on "Business and Human Rights in Latin America".

Gibbs, C., Gore, M. L., Mcgarrel, E. F., & Rivers, L. (2010). Introducing conservation criminology: Towards interdisciplinary scholarship on environmental crimes and risks. *British Journal of Criminology, 50*(1), 1–21.

Governo do Estado de Minas Gerais. (2017). *Comissão da Verdade em Minas Gerais: Relatório Final.* GOVEMG.

Goyes, D. R. (2019). *Southern green criminology: A science to end ecological discrimination.* Emerald.

Goyes, D. R. (2021). Environmental Crime in Latin America and Southern Green Criminology. *Oxford Research Encyclopedia of Criminology.* https://oxfordre.com/criminology/view/10.1093/acrefore/9780190264079.001.0001/acrefore-9780190264079-e-588

Grupo da Força-tarefa de Minas Gerais. (2016, February). *Avaliação dos efeitos e desdobramentos do rompimento da Barragem de Fundão em Mariana-MG.* Secretaria de Estado de Desenvolvimento Regional, Política Urbana e Gestão Metropolitana. https://www.agenciaminas.mg.gov.br/ckeditor_assets/attachments/770/relatorio_final_ft_03_02_2016_15h5min.pdf

Hall, M. (2013). *Victims of environmental harm: Rights, recognition and redress under national and international law.* Routledge.

Hamilton, M. (2021). *Environmental crime and restorative justice: Justice as meaningful involvement.* Palgrave Studies in Green Criminology.

Hamilton, M. (Forthcoming). *Restorative justice conferencing for environmental offending.*

HOMA. (2022a, February 14). A criação das assessorias técnicas para as comunidades atingidas. HOMA. http://homacdhe.com/index.php/2022/02/14/a-criacao-das-assessorias-tecnicas-para-as-comunidades-atingidas/

HOMA. (2022b, February 14). O que é a Repactuação? http://homacdhe.com/index.php/2022/02/14/o-que-e-a-repactuacao/

Johnstone, G., & Van Ness, D. W. (2007). The meaning of restorative justice. In G. Johnstone & D. Van Ness (Eds.), *Handbook of restorative justice* (pp. 5–23). Willan Publishing.

Lynch, M., & Long, M. (2022). Green criminology: Capitalism, green crime and justice, and environmental destruction. *Annual Review of Criminology, 5*, 255–276. https://doi.org/10.1146/annurev-criminol-030920-114647

Marshall, T. (1999). Restorative justice: An overview. *Home office research, development and statistics directorate.* http://www.antoniocasella.eu/restorative/Marshall_1999-b.pdf

Mazzola, B. (2018). *Da gestão dos stakeholders à licença social para operar: o caso do desastre de Mariana* [Doctoral thesis, Universidade de São Paulo—Faculdade de Economia, Administração e Contabilidade].

Mazzucato, C. (2018). Victims of corporate violence in the European Union: Challenges for criminal justice and potentials for European police. In G. Forti, C. Mazzucato, A. Visconti, & G. Stefania (Eds.), *Victims and corporations: Legal challenges and empirical findings* (pp. 58–59). Wolters Kluwer Italia.

MCold, P. (2001). Towards a mid-range theory of restorative criminal justice: A replay to the maximalist model. *Contemporary Justice Review*, *3*(4), 357–344.

Milaré, É. (2015). *Direito Ambiental* (10th ed.). Revista dos Tribunais.

Ministério Público Federal et al. (2022). Joint Press Release. https://www.mpf.mp.br/mg/sala-de-imprensa/docs/2022/caso-samarco/Pressrelease.pdf

Nunes, P. H. F. (2016). The amazon cooperation treaty organisation: A critical analysis of the reasons behind its creation and development. *Brazilian Journal of International Law, 13*(2), 220–246. https://www.corteidh.or.cr/tablas/r36669.pdf

Observatório Nacional. (2021). *3º Relatório Trimestral de Atividades Observatório Nacional sobre sobre Questões Ambientais, Econômicas e Sociais de Alta Complexidade e Grande Impacto e Repercussão 2020.* https://observatorionacional.cnj.jus.br/observatorionacional/wp-content/uploads/2020/11/3Relato%CC%81rio-de-Atividades-Observato%CC%81rio-Nacional_2020-1-2.pdf

Oliveira, C. R. (2021a). Facing problems of restorative justice in Portugal and Brazil: Learning with/from the fields. *Journal of Victimology, 11*(2), 95–118.

Oliveira, C. R. (2021b). *Justiça Restaurativa Aplicada. Estudo de caso das experiências do Brasil e de Portugal.* Blimunda.

Orlando, E., & Bergin, T. (2017). Forging a socio-legal approach to environmental harms. In E. Orlando & T. Bergin (Eds.), *Forging a socio-legal approach to environmental harm: Global perspectives.* Routledge.

Pali, B. (2016). *Doing restorative justice in intercultural contexts: An alternative discourse of justice and security* [Doctoral thesis, KU Leuven—Faculty of Law].

Prata, D. A. (2019). *Criminalidade Corporativa e Vitimização Ambiental: análise do caso Samarco.* LiberArs.

Prata, D. A. (2020). Corporate crime and environmental victimisation: Analysis of the Samarco case. *International Review of Penal Law, 91*(1), 203–223.

Prata, D. A., & Garrido, D. B. (2022). Brazil: A progressive framework on civil liability and human rights protection? In E. Aristova & U. Grusic (Eds.), *Civil remedies and human rights in flux: Key Legal developments in selected jurisdictions* (pp. 87–107). Hart Publishing.

Quijano, A. (2005). Colonialidade do poder, eurocentrismo e América Latina. In L. Edgardo (Ed.), *A colonialidade do saber: eurocentrismo e ciências sociais—perspectivas latino-americanas* (pp. 107–30). Consejo Latinoamericano de Ciencias Sociales (CLACSO).

Ramos, G. C. D. (2010). *América Latina y el Caribe como reservas estratégicas de minerales. Ecología política de la minería en América Latina: aspectos socioeconómicos, legales y ambientales de la mega minería* (pp. 17–57). Centro de Investigaciones Interdisciplinarias en Ciencias y Humanidades Universidad Nacional Autónoma de México.

Rosenberg, M. (2006). *Comunicação Não-Violenta.* Técnicas para aprimoramentos pessoais e profissionais. Ágora.

Saad-Diniz, E. (2019). *Vitimologia Corporativa.* Business & Criminal Justice. Tirant Lo Blanch.

Saad-Diniz, E., & Gianecchini, J. V. (2021). The gears of environmental victimisation. *Revue Internationale de Droit Pénal*, *9*(2), 347–374.

United Nations Office on Drugs and Crime—UNODC. (2020). *Handbook on restorative justice programmes. Criminal justice handbooks series.* United Nations. https://www.unodc.org/documents/justice-and-prison-reform/20-01146_Handbook_on_Restorative_Justice_Programmes.pdf

Varona, G. (2019). Restorative pathways after mass environmental victimisation: Walking in the landscapes of past ecocides. *Oñati Socio-Legal Series, 10*(3), 664–685.

Vera, N., & Vargas, L. (2021). *Formación social y deforestación en la Amazonía boliviana. Expansión mercantil capitalista y la Amazonía como nueva frontera de recursos en el siglo XXI.* Consejo Latinoamericano de Ciencias Sociales (CLACSO).

Verchick, R. R. M. (2010). *Facing catastrophe: Environmental action for a post-Katrina world*. Harvard University Press.

White, R. (2008). *Crimes against nature: Environmental criminology and ecological justice*. Willan Publisher.

White, R. (2013). Environmental Harm An eco-justice perspective. Policy Press University of Bristol. https://www.lpwntb.or.id/wp-content/uploads/2021/03/Policy-Press-Studies-in-Social-Harm-Rob-White-Environmental-Harm_-An-Eco-justice-Perspective-Policy-Press-2013.pdf

White, R. (2014). Eco-justice and problem-solving approaches to environmental crime and victimization. In T. Spapens, R. White, & M. Kluin (Eds.), *Environmental crime and its victims: Perspectives within green criminology* (pp. 87–102). Ashgate.

Wolf, B. (2011). "Green-collar crime": Environmental crime and justice in the sociological perspective. *Sociology Compass, 5*(7), 499–511.

Zehr, H. (2008). *Trocando as lentes. Justiça Restaurativa para o nosso tempo.* Palas Athenas.

9

Beyond Retributive Justice: Listening to Environmental Victims' Demands in Brazil

Marília de Nardin Budó, Karine Ágatha França, and Lorenzo Natali

Introduction

"Green criminology" is a theoretical perspective oriented towards opening criminological paradigms to issues of environmental harms and crimes. It allows for diverse theoretical orientations to dialogue with the aim of connecting and analysing crimes, harms, and injustices related to the environment—including animals, vegetal forms of life, ecosystems, the planet, and the biosphere (Lynch & Stretesky, 2003; Natali, 2013; South, 1998, 2014; South et al., 2013, p. 28; White, 2011).

The complexity of environmental crimes and harms, especially those with massive effects and caused by transnational corporations with state

M. de Nardin Budó (✉) · K. Á. França
Federal University of Santa Catarina (UFSC), Florianópolis, Brazil
e-mail: mariliadb@yahoo.com.br

L. Natali
University of Milano-Bicocca, Milan, Italy

D. R. Goyes (ed.), *Green Crime in the Global South*, Palgrave Studies in Green Criminology, https://doi.org/10.1007/978-3-031-27754-2_9

collusion, challenges the traditional socio-criminological categories and, more generally, compels a profound rethinking of the idea of modernity itself (see also South, 1998). Green criminology, in particular, tries to overcome the inability of mainstream criminological knowledge to ask the decisive question of how the social sphere and the natural sphere are interwoven, transcending some dichotomies of modern thought—such as those between persons and things, between technology and society, between culture and nature, and, last but not least, between retribution and reparation (Halsey, 2006). In this regard, green criminology suggests that we need new conceptual tools and also different ways to respond to the tragic consequences of the human–environment relationship.

To deal with the theme of environmental harm and justice in Latin America, we adopt a decolonial perspective, suggesting that modern rationality is also a colonial (and neo-colonial) rationality (Quijano, 2000). In this sense, this chapter falls under the conceptual umbrella called by Goyes (2019, p. 11) a Southern green criminology, "as the science that is attentive to the dynamics and contexts of the Global South and grows out of the epistemological power of the marginalised, impoverished and oppressed".

Southern green criminology is based on the epistemological criticisms of modern science, and mainstream criminology. "Coloniality of knowledge" (Quijano, 2000) is one of the most important categories in this approach, as it denounces how the centrality of the global North in the production of science methods and theories has left for the global South the role of being a mere source of data and a *reproducer* of those methodologies and theories (Carrington et al., 2016; Connell, 2007). Science produced in the global North can travel in time and distance, as universal knowledge; science made in the global South is seen as local, specific, and exotic knowledge (Budó, 2021; Quijano, 2000; Santos, 2016).

Moreover, this approach highlights how racism and androcentrism are the core of modern science, creating a hierarchisation of knowledge and, at the same time, of bodies. Male white people from the global North are the reference point from where all gender, racial, and ethnic dimensions are to be measured, examined, and analysed (Bonilla-Silva, 2012; Harding, 1993). In this sense, knowledge produced by such men will be considered universally valid while knowledge produced by "others" is considered only relatively valid or just invalid. This is why one of the consequences of the coloniality of knowledge is "epistemicide" (Santos, 2016, p. 238), "the massive waste of social experience and, particularly, in the massive destruction of ways of knowing that did not fit the dominant epistemological canon". The resistance to this process is called by Santos "the epistemologies of the South", as ways of knowing that are born in the struggle against capitalism, colonialism, and patriarchy.

Furthermore, being under this epistemological umbrella compels Southern green criminology to move beyond the racism characteristic of modern science and to promote a different way of understanding and interpreting knowledge. As Santos (2016) argues, the colonial violence murdered the rival knowledges deemed to be non-scientific—knowledges that now call for reparation (see also Topçu et al., 2022): "As a result, there is no global social justice without global cognitive justice" (Santos, 2016, p. 296). Therefore, the claims for justice coming from environmental victims are central to Southern green criminology. This chapter fits this proposal, as listening to and taking into account the knowledge produced by people affected by environmental harm can reveal different meanings of justice to deal with environmental victimisation.

We seek to show how people affected by environmental harm co-construct the meanings of harm and justice, especially regarding the activities of massive corporations in Brazil. From a review of the empirical research with environmental victims, we develop categories of analysis to theorise how the principles of restorative justice can dialogue in the construction of alternative perspectives of environmental justice.

Environmental Crime, Harm, and Injustice

The definition of crime has been an epistemological problem to criminology since its origin. After the "destructuring impulse" (Cohen, 1985) brought to the field by critical criminology, the problem moved to another place: how to deal with the problematic situations that actually happen in the world, even if we do not use the word "crime"? Or to use Baratta's words (1985), which would be the "material referent" of criminology?

To deal with this question, in this chapter we adopt the social harm approach (Davies et al., 2021; Hillyard & Tombs, 2004) applied to the environmental field. Firstly, because the state is one of the major perpetrators of environmental harms, it is not interested in defining its own harmful actions and omissions as crimes in the legal sense (Halsey, 2004, p. 836; White, 2008; 2011, p. 6). Secondly, because even when the private sector causes environmental harm, the companies in general have privileged access to parliament and to the justice system. It is almost impossible to assign liability in these cases (Whyte, 2020). Moreover, it is because environmental crimes are placed along the legal–illegal *continuum* (including both criminal and lawful actions) that criminological analysis should not stop at behaviours forbidden by the law, as Edwin Sutherland already reminded us in the 1940s (Ruggiero & South, 2013).

At the same time, a political economic approach to environmental harm shows how the functioning of the criminal justice system perpetrates ecological destruction, since it also functions for capitalism (Baratta, 2002; Budó & Falavigno, 2020). In this sense, it is also important to expose the structural delegitimisation of the penal system to prevent and repair harms, moving beyond the illusions promoted by modern penal rationality (Andrade, 2003; Pires, 1999). The contemporary discourses about criminal law reform are still fundamentally marked by certain knowledge and conceptions of modernity, which are cognitive obstacles to innovation. As a system of thought, modern penal rationality prevents us from moving beyond a traditional idea of justice (Pires, 1999, p. 65).

As Alvaro Pires explains, modern penal rationality is part of a dichotomous way of interpreting law. On one hand, civil law is oriented to the domestic and private world or to the private interests of plaintiffs and defendants; on the other hand, criminal law deals with the public world and collective interests, excluding the interests of victims and defendants:

> The first one is imaginative, flexible and creative: it listens to the plaintiff, accepts negotiation and mediation between the parties, shares what is mine and what is yours searching for justice and equity, and it can impose reparation and invoke different kinds of prescriptions. The second is the

> poor cousin of the family: badly equipped and limited from the point of view of the possible decisions to solve the conflicts, it has the simple function of distributing the evil and of paying in the same coin. (Pires, 1999, p. 66)

Therefore, searching for an alternative criminal policy able to solve conflicts beyond punishment necessarily demands a different *imagination* about what justice is.

In this regard, it is necessary to underline that the different definitions of justice provide different theoretical foundations to what we think of as "harmful" and/or "criminal" from a Southern and ecocentric perspective. And, furthermore, such definitions allow the articulation of concepts relating to *who* can be considered a "victim". This in turn allows us to reflect more specifically on what "repairing" such harm might look like (see Natali, 2021; Natali & Hall, 2022).

In this chapter we address specifically aspects of *environmental justice* (see also White, 2011) as our main focus is on the inequalities and oppressions that naturalise and make invisible the sufferings of the people who are most affected by environmental harm. At the same time, looking at the struggle against environmental injustice, we suggest that the starting point for restorative justice practices in this field *should* be the *listening* to the victims' claims and narratives (Pemberton et al., 2007, 2019).

Environmental Victims

Rather like the victims of white-collar crimes, environmental victims often remain hidden in the shadows (Davies et al., 2014; White, 2011, p. 109). More generally, environmental victimisation poses a series of new questions that criminal justice systems are unprepared to face (Hall, 2013, pp. 219–220). First, the harms can be suffered by an extended group of people (sometimes with conflicting interests), by nature, and by animals other than humans (Varona, 2022; White, 2008; Wyatt, 2022). Second, the perpetrators are often corporations or states (White, 2011, pp. 103–104; Whyte, 2020). Finally, the causality nexus is extremely complex to reconstruct, sometimes leading to environmentally harmful acts being considered "crimes without victims" (see Bisschop & Vande Walle, 2013, p. 40).

The relevant scientific literature shows how the difficulties encountered in establishing a causal relationship between the polluting activities and the contaminating effects may offer perpetrators an easy escape route (Williams, 1996). For example, the various strategies of neutralisation of responsibility on the part of corporations or the state include: denying the problem (Budó, 2021; Cohen, 2001); putting into perspective what is now seen as damaging but that could be interpreted differently in the future; justifying the harms of economic development, being so-called "sustainable development", one of the main greenwashing strategies (Galtung, 1990; Krenak, 2019); and reproaching, blaming, dividing, and confusing the victims (Williams, 1996). For all these reasons, it is important to explore the nature of environmental victimisation as an active social process, which implies relationships of power, control, and resistance (White, 2011, p. 106).

Moreover, scholars in this field suggest that it is necessary to understand how gender and race relationships as well as human–animal relationships influence the way in which victimisation is perceived (see also Sollund, 2022, p. 300). Regarding racism, it is easy to see how there is an objectification of non-white bodies, considered less valuable than white bodies in different societies, which is also called environmental racism (Pulido, 2018). This happens because the former is seen in these androcentric, anthropocentric, and racist structures as "the other". This otherness leads to the denial of the harms produced and engenders processes of normalisation and invisibilisation that lays at the core of the coloniality of power (Quijano, 2000).

In this sense, North–South relationships also show how racism in a broad sense can naturalise, in Southern territories, harms that would be unacceptable in Northern ones. Double standards of safety, collusion between big transnational corporations, and ancient local oligarchies are just some of the strategies used to reproduce structural violence through environmental victimisation (Böhm, 2019; Galtung, 1969). In this sense, thinking about justice also means dealing with all types of violence that are behind specific cases of environmental harm (Silva Neto, 2021).

Moreover, given the still limited research on the impacts of environmental harm, it is difficult to paint a full picture of these impacts. Particularly concerning is the lack of research asking (human) environmental victims *themselves* about their experiences and support needs. We still lack detailed data on the lives of people living in polluted areas, describing *from their perspective* what they know of, think about, and feel towards the reality in which they live (Natali, 2016). Such an orientation raises questions regarding how people live and make sense of their experiences in polluted places or in cases of massive environmental tragedies, such as the ones we discuss in the second part of this chapter. Empirical research on the inhabitants of highly polluted places has revealed the multiple interpretations guiding their decision to act individually or collectively (or to abstain from acting) in response to their environmental victimisation (Allen, 2018; Auyero & Swistun, 2009, p. 159). Moreover, social and cultural perspectives on environmental victimisation processes have proved extremely important in showing that experiences of environmental suffering are not simply individual: they are collectively created in the wider social macrocosmos where the residents are situated and situate themselves.

Environmental Suffering and Experiences of Injustice

What is clear from the literature is that the impacts of environmental victimisation are extremely varied and wide-ranging, a main reason why states have difficulty in implementing proscribed or firm procedures of regulation and restitution. Through the visibility of the victims' narratives and demands, it is in the interest of the green criminological perspective to think and collectively build possibilities that best respond to the scenarios of socio-environmental violence (Budó, 2019; Colognese, 2018; França, 2022; Natali & De Nardin Budó, 2018; Silveira, 2018).

One of the strategies adopted by the powerful to avoid any liability for the harms they perpetrate is silencing peripheral voices. With state collusion, powerful companies perpetuate harmful activities discreetly, invisibly, and sophisticatedly (Cruz et al., 2020). Imagining justice from the voices of people harmed by these activities makes it possible to understand the dimensions of suffering, which permeate the multiple realities of violence experienced, and the claims, demands, and struggles promoted. Green criminology, guided by a Southern perspective, is dedicated to issues related not only to legal definitions of crime, but also to victims' political claims and mobilisations (Hall & Varona, 2018).

Methodologically, approaching experiences of environmental victimisation is always a challenge. Scientists have a position in the world and all knowledge developed by science is situated therein (Haraway, 1988). Even the concept of victim and the process of environmental victimisation is an area of political controversy. Many activist groups that deal with environmental harm in communities sometimes use other words to describe their identity. For example, "affected" (Movimento dos Atingidos por Barragens—MAB, Brazil) and "exposed" (Associação Brasileira dos Expostos ao Amianto—ABREA, Brazil) (see also Natali, Suman, Budó, 2023). Being part of environmental justice movements and actively listening to the claims of the people who have been harmed also means being open to completely new ways of comprehending the experience of victimisation.

Furthermore, we know that the people most affected by environmental harms are women, non-whites, and Southern people (Goyes, 2019; Pulido, 2018; WHO, 2010). At a global level, this is expressed, for example, by the transfer of harms from the global North to the global South, the double standards of safety and regulation used by transnational companies (Böhm, 2019), and the disputes over territory in which these companies sometimes engage (Whyte, 2020). Racism, androcentrism, and coloniality are structures of oppression that differentiate the people deserving a healthy life from the disposable ones (Mbembe, 2003). In this sense, from an androcentric, white, and Northern science perspective, environmental victims are "the other", the "subaltern". When we talk about the harm caused by the coloniality of knowledge, this is also about how "white discourse" can make the harm and suffering of most of the environmental victims invisible. As they are seen as objects, their voices are not important enough to be listened to. To overcome coloniality and to promote a pathway to alternative forms of justice, we must ask, paraphrasing Spivak (1994): Can the environmental victim, as a subaltern, speak?

To address this question, we selected the results of a few empirical studies focusing on the victims of environmental crimes and their feelings about justice and injustice. The studies were developed mostly in Brazil and they describe different cases of exposure to environmental harms. In these four cases, we consider the data obtained in each work and identify the main points that connect the perspective of justice for the victims. The cases are listed and synthesised as follows.

1. *The asbestos case* (Budó, 2019; Natali & Budó, 2018; Silveira, 2018): asbestos is a flexible, incombustible, indestructible, cheap mineral that was widely mined and used as a raw material in industry in the twentieth century. Different types of asbestos are found in nature, but nowadays all types have been banned in 60 countries because of its hazardous effects on human health, although in most Southern countries it is still mined and used. The World Health Organization calculates that at least 107,000 people die every year in the world because of asbestos related diseases (ARD), caused by the inhalation of small particles of the fibre. Furuya et al. (2018) claim that this number is an underestimation and suggest that 255,000 deaths annually is nearer the mark. The studies approached here were developed through in-depth interviews with people affected by the asbestos harms spread by the company Eternit in Osasco (SP), in Brazil, and in Casale Monferrato, in Italy.
2. *Dam collapses in Mariana (2015) and Brumadinho (2019)* (MG-Brazil) (Colognese, 2018; França, 2022): On 5 November 2015, the Fundão Dam, located in the city of Mariana, Minas Gerais, Brazil and explored by Samarco (a joint venture of Vale and BHP Billiton), collapsed. The dam spilled 55 million cubic metres of iron ore tailings. The dust travelled more than 600 km along the river Doce, bringing everything along with it—people, animals, houses, cars—into the ocean, provoking still inestimable socio-environmental

harms; 17 people were killed; 85% of the buildings in the localities were completely destroyed. On 25 January 2019, the dam of the Mine Córrego do Feijão I, in Brumadinho (MG), Brazil, collapsed, and a great avalanche of sludge travelled 9 km with 12.7 million cubic metres of mining tailings. The mud engulfed buildings, construction sites, houses, and also reached the Paraopeba River. The damage caused by the rupture constituted the largest occupational accident ever recorded in Brazil, accounting for 246 deaths, among these 120 workers, employees of the company Vale S/A, as well as outsourced employees (Reis, 2019). There are also 24 missing persons, which according to experts would bring the total to 270 deaths (Senado, 2019). The case is being considered by some scholars as the second deadliest global industrial accident of the twenty-first century (Freitas & Silva, 2019). The research methodology consisted, firstly, in field research carried out in the territory of Mariana (MG), in 2018, where 12 victims of the failure of the Fundão dam of the company Vale/SA were interviewed (Colognese, 2018). Secondly, during the pandemic period, in 2021, five affected people from Brumadinho (MG) were interviewed in virtual mode (França, 2022).

3. *The Itá hydroelectric dam* (SC-Brazil) (Cruz et al., 2020): This is one of the plants in the Uruguay River basin and was built in the 2000s, was responsible for the environmental victimisation of more than 3560 families, having forcibly displaced 827 of them, in addition to causing a series of devastating consequences for local communities. The construction of this dam impacted 11 municipalities: Aratiba, Mariano Moro, Severiano de Almeida, and Marcelino Ramos (in Rio Grande do Sul [RS]), and Itá, Arabutã, Concórdia, Alto Bela Vista, Ipira, Piratuba, and Peritiba (in Santa Catarina [SC]). The methodology adopted by the authors was based on participant observation, as well as narrative interviews with 12 participants. Data collection took place in 2016, in several communities affected by the Itá hydroelectric plant, in Santa Catarina.
4. *Belo Monte hydroelectric dam* (PA-Brazil) (Ebus & Kuijpers, 2016): The third largest plant in the world, installed in Brazil, was designed in 1975, but construction was stopped due to widespread resistance from local, regional, national, and even international communities.

> Building resumed under the 2001 government and the dam became operational in 2016 (Fainguelernt, 2020; Farias, 2014). Belo Monte is responsible for the displacement of 50,000 people, an exponential increase in violence, prostitution, drug trafficking, diseases, public health crisis, pollution, deforestation, and changes in the river flow (Rosa, 2016).

All these studies address a debate about justice from the victims' perspective. At the same time, they approach very different processes of victimisation. For example, in the asbestos case, the diseases suffered are all long term. Thus, decades passed between the exposure to asbestos and the awareness of the harms caused by it. In the two cases of dam collapse, the harm can be seen in seconds. So, we can say that these are short-term experiences of victimisation. Simultaneously, the context of the hegemony of the mining industry in the territory of Minas Gerais also entails a long-term exposure to environmental harm, as, for example, the contamination of water, soil, and air, the violence perpetrated against Indigenous people to take their territory, and the decrease of fish in the river over the years. In this sense, from a microanalysis of the field to a macroanalysis, we can say that all of them have, as a background, suffered nature and human exploitation by big national and transnational companies.

Empirical research that contains environmental victims' participation and testimony allows the subversion of the destructive order that orientates the interests of powerful actors, especially when it links academic knowledge and the praxis of the social movements and social organisations. In these contexts, the mobilisation of environmental victims ends up being part of a process marked by conflicts, insubordination, and insurgencies that promote a transformation in the public space, strengthening an alternative rationality that can be well developed in the decolonial approach to the theory of the commons (see Federici, 2009; Gago, 2019).

Reparation and Accountability: The Struggle for Rights and Recognition

The *feeling of abandonment and neglect* over the suffering of people affected by environmental harm commonly appears in the victims' narratives. In the case of asbestos in Osasco (SP), the same company that perpetrated the harm was the employer of the victims. This context brings important complexities to the debate about justice because the strategies used by the company to deny the harms were characterised by disinformation and cooptation (Silveira & Budó, 2021). For example, one of the participants of Silveira's study narrates that the families of the workers harmed had the support of the company only as long as the work relationship existed (Silveira, 2018, p. 90), but the harm suffered by the victims does not cease with the termination of this relationship. It should be acknowledged that the "ties" binding the perpetrator of the harm to the victims should not be determined by an "employment contract". More importantly, the relationship between perpetrators and victims should not be terminated while the latter and their families continue to suffer consequences without any form of reparation.

The *lack of acknowledgement* is not only related to the company. In Silveira's study we can also find this feeling of abandonment regarding the state, revealing the *lack of trust and credibility about the formal and institutional instances of justice*. Many of the participants in the research identified omissions on the part of the state and one of them actually accused the state of actively "working on behalf of the bosses. Unfortunately, the power of the state is used for the bosses!" (Silveira, 2018, p. 188). This perception of the power relations between capital and work turns into disobedience, as that feeling of abandonment becomes a collective struggle. The opponents in this case are not only the direct perpetrators of harm but the state itself. In their study on the flooding for the construction of the Belo Monte hydroelectric dam, Ebus and Kuijpers (2016) also address the lack of confidence in the judicial system. The authors interviewed the leaders of the people affected, and the narratives denounced the collusion between judges, the company Norte Energia, and the inspection organs of the state. "The current Judiciary in Brazil behave in consonance with corporative interests. The judicial system does not really investigate and sue in the cases opened by the prosecutors, ... most of the judges are instruments of the State to accomplish their goals of economic development" (Ebus & Kuijpers, 2016, p. 141).

At the same time, the *incommensurability of the perpetrated harms* in all these cases appears to be a serious obstacle to determining what kind of justice would be possible. For example, in the case of the dam collapse in Brumadinho, one of the participants, whose sister was dragged to her death by the sludge of iron ore tailings, narrates that "everything is reparable, but human loss is irreparable, there is not a possible value you can attribute to it. There is not possible indemnity ... it is ... we have to live with this absence, with this lack, for the rest of our lives" (França, 2022, p. 197). The collective pain feels like a total loss to her: "It is much more than an environment destroyed, they destroyed lives, an entire city, it is a part of Minas that is still bleeding. A part of Minas Gerais is marked like this, ... there is no turning back". The sense of time in this case has a particular importance because the destruction *narrated* by the participant occurred in seconds. We can see images of the security cameras and in one second the landscape is destroyed. The temporality is short, and the sense of "no turning back" seems to be stronger than in the other cases (see also Chateauraynaud & Debaz, 2022).

Both in Brumadinho and Itá the victims also highlighted the importance of economic indemnity and the necessity to *repair* the damage perpetrated. In Itá, the research by Cruz et al. (2020, p. 8) reveals how a person affected talks about the feeling of abandonment and neglect on the part of the company after environmental victimisation: "they went there to my house, took my ID and other documents to see if they could get something back yet. To this day nobody else came" (Cruz et al., 2020, p. 8).

Because of this indifference, the participants narrated that a huge struggle was needed to obtain reparation. With the aim of achieving such reparation, the organisation created by those affected sabotaged machinery, blocked roads and bridges, as well as staging riots and protests. As a result, they had to resist repression on the part of the state aimed at fragmenting people and annihilating the possibility of liability. One of the participants said that "I was sued six times, because from the moment they suspected that we were going to mobilize, a prohibitory interdict came already communicating that if we mobilized, if we did any action that harmed them, we would be prosecuted" (Cruz et al., 2020, p. 9). In this sense, the system of justice operated in the opposite direction: *instead of acknowledging that those people were victims of a state–corporate crime, they were accused of being themselves perpetrators of crimes.* These are also important issues to be studied in Southern green criminology: the selectivity of the criminal justice system in environmental crimes (Weinstock, 2017; Weis et al., 2022) and the criminalisation of social movements.

In the face of this massive loss, the traditional system of justice becomes still more inadequate. In the cases studied, the victims' perceptions that the "justice of humans" cannot be proportional and fair, also reveal a hope in a "divine justice" as the only possibility for them to be listened to and to obtain a response. Silveira (2018, p. 98) says that "even if they mentioned the desire of receiving monetary indemnity and payments, of seeing jail convictions and the total ban of asbestos and asbestos tailings in a national level, the divine justice appears as a more effective kind of justice for the participants". At the same time, we cannot really know if this appeal to religion would be the same if the people affected by the environmental harm were listened to in their demands and claims in alternative practices of justice (see also Globus et al., 2014).

In some of these studies, the participants' perceptions of justice seem to be far from the idea of vengeance or hate and dehumanisation. They wanted "justice to be done". In the Brumadinho case, one of the participants said that "from the humans' justice we hope that it works as an example … an exemplary punishment so that it doesn't happen again" (França, 2022, p. 193). One of the participants of Silveira's study on the asbestos harm in Osasco (SP) also expressed a sense of justice that moves more to the future than to the past. The achievement of the asbestos ban in the Supreme Court in 2017 was a consequence of the collective struggle of ABREA and it appears in the participant's narrative as a strong victory: "It's been gratifying! The people who are recurring to civil suits nowadays are achieving something! Just the fact that the Supreme Court banned asbestos for us is already a victory" (Silveira, 2018, p. 96).

Keeping the memory alive through testimony challenges the wide denials of the project of forgetfulness and erasure of the responsibilities of powerful agents. This project is developed by reducing the forces and capabilities of the political organisation of the victims, who seek to make visible the social and environmental injustices through insurgent actions (Gago, 2019; Silva Filho, 2012; Varona, 2020). The link between the trauma from the past and the testimony in the present is a pathway to avoid denial and the repetition of harmful behaviours by the state and the companies in the future. The importance of the victims' testimony in cases of massive violations of human and environmental rights is not only about healing, but it is also about constructing a collective memory and identity to challenge the operations of greenwashing and other strategies of denial that are used by the powerful to achieve forgetfulness.

The demand from the victims also lies in the search for public recognition of the harm caused. This includes the participation of civil society in the construction of public spaces for debate, listening, and welcoming. The category of cultural trauma is interesting from the point of view of the effectiveness of justice in contexts of victimisation, since solidarity and social responsibility enables the sharing of pain and suffering caused to the victims. These processes contribute to the fight against forgetfulness and silencing produced by the powerful (Goyes & Franko, 2022; Seligmann-Silva, 2008). Thus, social responsibility includes awareness about history, and the cultural, racial, ethnic, and gender inequalities that form the relationships of each social context, in order to recognise that the social and environmental trauma the victims are suffering is a result of the colonial, racist, patriarchal, and speciesist heritage.

Without testimony as a tool to ensure memory, there is no possibility of bringing the present closer to the events of the past, nor of expanding the relationship between victim/survivors and society. When spaces for testimony are lacking, negationism becomes present and takes a leading role in relation to survivors, generating a double victimisation. It also destroys the possibility of the social survival of those affected and the connection of past traumas with society (Piralian, 2000). Therefore, the link between individual memory and collective memory is broken (Goyes et al., 2022). By recognizing the vulnerabilities of the other from our own vulnerabilities (Butler, 2011), it becomes possible to avoid further revictimisation and, consequently, the reinforcement of silencing and the erasure of harm and the victims in these contexts (Ginzburg, 2001).

Truth and Memory: Opposing the Forgetfulness Project

> A jewel is a thing, it is putting a price ... nowadays I understand that this jewel is in the sense of valuing … it is more than giving a price, it's giving value, and these people really were jewels, they had an inestimable value to each family, right? (França, 2022, p. 194)

In all cases discussed in this chapter, the need of the environmental victims to tell their stories from their own perspective has remained clear. More than that: the claim is also to be listened to, as there is a huge difference between the access to public discourse by the state and the companies and by the local people affected by environmental harm.

The project of silence and amnesia causes over-victimisation. The victims are relegated to a place of escape and a freezing of the self in relation to the environmental victimisation suffered (Franko & Goyes, 2023). Therefore, testimony offers a possibility of promoting healing from the harmful events, through their memories and participation, by sharing their painful experiences with civil society: "Victims who choose to tell constantly seek a voice and a narrative that allows them to make sense of what happened and honor what they lost" (Franko & Goyes, 2023, p. 49).

In França's study, one of the participants told her that to speak with researchers is also a way to publicise the truth "to the people to understand that what Vale say isn't the reality, isn't the truth" (França, 2022, p. 182). The participant shows that *the victims' discourse is a counter-discourse, because the "official" discourse is the one told by the transnational company*. She also declared that her intention was not to damage the company's image, but "to deny the lies Vale is telling" (França, 2022, p. 141). In this perception, the participant recognises the importance of contradicting the "official discourse", but at the same time recognising the importance of the company to the local economy. The feeling seems to be also ambiguous, neither of resignation nor of anger.

Another participant of França's study said something similar, dealing with the importance of testimony and the fear of repetition: "I wish that this will never happen again, that nobody else has to go through this, ... that Vale learns something you know? That everybody learns something" (França, 2022, p. 182). The investigation and the public knowledge about the events are part of the construction of truth for the people who went through a traumatic episode. In this sense, identifying and acknowledging the responsibilities, and making visible their consequences, are important for social, political, and educational binding for the construction of mechanisms that also strengthen democratic spaces.

In the discourse of the environmental victims, the collective reconstruction of the happenings of the past is a mechanism to disrupt the project of forgetfulness promoted by the companies and by the state. In the case of the Vale dam collapse in Brumadinho, the participants also said that the official instruction given to the people harmed was to forget the traumatic episode and just move on with their lives. In contrast to this policy, the people harmed have mobilised on the 25th of each month since January 2019, so that nobody would forget the tragedy. One of the participants also said, "I am leaving this planet in some years, but I think that this part of the history will be recorded so the people don't forget that what happened could have been avoided" (França, 2022, p. 192). This testimony denotes the confidence in the transgenerational awareness about the injustices perpetrated in the past, keeping them alive so that justice can be materialised (Mate, 2015). To use Le Ven's (2014, p. 24) expression, "don't forget, so that it never happens again".

The *struggles for truth and memory* are also present in the discourses of the people affected by asbestos in Osasco. Silveira shows that the motivation for the struggle for the asbestos ban was also a result of the struggle to keep the memory alive. As one of the participants says, the anger resulting from the enormous loss of friends and relatives is strong, but the knowledge that something was done to stop this harmful industry is still stronger: "To keep only thinking about the anger is worse and it's not going to solve anything, and it's going to hurt ourselves even more. So, let's fight against this disgrace to stop it and to stop also other products as benzene, mercury, the dust that causes harm to the comrades from Santos. We are linked to all these struggles nowadays" (Silveira, 2018, p. 91).

Some of the narratives also showed how the memory of the tragedy in the case of the dam collapse in Brumadinho was represented in *objects* that recall that day. One of the participants said that she keeps her boots dirty with mud. Another participant said that he keeps his dirty clothes that have dust on them from the knees down. Forgetfulness, silence, and the erasure of memory reinforce the naturalisation and banalisation of the history of the deaths and the survival, provoking a collective catastrophe (Seligmann-Silva, 2008). For this reason, testimony is an instrument to preserve memory, allowing that the present and the past remain close, to be comprehended not only from a victim–survival perspective, but including others that can be part of this listening (Ginzburg, 2001; see also Gorostiza & Armiero, 2021).

The connection between the memory socially constructed by the citizens of the destroyed places and justice also appears in the narrative of a participant of Brumadinho: "It appeared in the first page of the journals, not only in Brumadinho, but in the whole world. Nowadays this is not remembered, it's not important to the people" (França, 2022, p. 193). Without memory and acknowledgement of the past it is not possible to face the present with responsibility, considering that responsibility comes from a historical conscience (Mate, 2015).

As a way to promote memory and truth for the victims of the Brumadinho dam collapse, one of the participants said that they started to call the people who died "jewels" because the former president of the Company Vale, Fábio Schvartsman, said in a public speech that "Vale was a jewel", without standing and proposing a minute of silence as a sign of respect for the people harmed. The expression was incorporated in the victims' counter-discourse, by disputing the narrative being played: "We started saying that Vale was not a jewel, that jewels were our beloved ones that it killed" (França, 2022, p. 206).

Those narratives make it clear that the relatives and survivors of these tragedies are trying to resignify the past through actions in the present. The memory policies are as important as financial compensation, making the sense of reparation reach another level to discuss justice. The environmental harms and injustices, as well as the responsibility of the powerful agents that provoked those harms, are, in this sense, publicly acknowledged (Varona, 2020). What to remember and for whom to remember are important questions to be problematised while we think about the uses of memory (Borges, 2016, p. 9).

Reflecting on the Victims' Claims for Justice from a Southern Perspective: An Open Conclusion on Restorative Justice

In the perspectives and narratives of environmental victims collected in the studies mentioned, we can see a deep relationship between memory and alternative perceptions about justice that are compatible with some of the principles of (environmental) restorative justice: (1) the lack of credibility in state institutions by the victims; (2) the search for alternative ways to make the perpetrators accountable, mostly based on the idea of reparation; (3) the idea that reparation is not only perceived as a monetary indemnity, as it is impossible to give a price to life, but also as a way to avoid the harm happening again; (4) the idea that reparation is thought of as a process, in which the conflict can be made explicit through the victims' testimony, to avoid their perceptions and demands being erased and silenced; (5) the importance of collectivity in this search for justice, understanding that it is possible to construct bridges through dialogue between victims, institutions, and civil society as a means to promote transgenerational responsibility, avoiding harm in the future.

The idea of reparation can include a diversity of methods and instruments so as to promote the public acknowledgement of social and ecological injustices and the accountability of the agents. The strategies to achieve these goals can be organised inside the system of justice, or they can be independent and autonomous from judicial intervention (Varona, 2020).

To address environmentally destructive activities by listening to environmental victims and respecting the complexity of environmental harm, we could draw inspiration from the criminologists' considerable interest in restorative justice mechanisms (Aertsen, 2018; Natali & Hall, 2022; Pali & Aertsen, 2021; Pali et al., 2022). First of all, Southern epistemologies bring the need for challenging the hierarchy of knowledge and underline the importance of understanding conflict resolution strategies that are already used by local communities. As Orth et al. (2020) note, bringing a Southern perspective to restorative justice is to focus on local needs with the potentiality of social emancipation, recognising and honouring the knowledge of Indigenous peoples, and other kinds of conceptions about conflict resolution that are not considered from a modern Northern rationality. At the same time, cases where the magnitude is great, such as the ones we have described, bring huge challenges for restorative justice. Coloniality seems to be the greater problem when dealing with these significant harms, as these ecological tragedies and victimisations are inscribed by racism. At the same time, anthropocentrism cognitively makes it harder to understand the meaning and extension of the harm perpetrated.

Restorative justice cannot be expected to be effective in all, or even many, cases of environmental crimes; nonetheless, it can contribute to repairing the resulting harms of environmental victimisation, having in mind the complexity of the definition of victim, perpetrator, and even harm in cases like the ones described (Budó & Pali, 2023). That is why, in the environmental field, it seems necessary to imagine a different system of justice: this does not mean aspiring to a utopia but, rather, imaginatively orienting towards possible futures, utilising languages and models of justice "other" than the traditional ones. Restorative justice—based on the idea of spontaneous participation and aimed at constructing, together and actively, a work of reparation for an act's destructive effects—represents a possible horizon far from any all-encompassing ambition (see also White, 2017).

Restorative justice has recently appeared in the current scientific literature on corporate violence as an alternative to promote the accountability of public and private agents that produce harm (Aertsen, 2018; Pali et al., 2022; Saad-Diniz, 2019). The victims' demands are at the core of this approach, and in cases of corporate violence it means recognition, protection, information, and support (Aertsen, 2018, p. 239). In Latin America, we also have to deal with the threats that victims and environmental activists are always receiving when they challenge the violent practices of the state and corporations.

In any case, to deal with the reality of environmental crimes and harms, it is also important to understand the colonial and racial dimensions of the knowledge produced by the powerful to address ecological destruction (Rodríguez, 2022). Also, the power imbalance between the stakeholders in most of the cases (Aertsen, 2018) means that protection should be offered, both to prevent retaliation against the victims by the company, and future exposure to harm, to the same and other victims.

The principles and values of restorative justice are defined by Aertsen (2018, p. 244) as follows:

> (a) Restorative justice is a model of doing justice that is rooted in an immediate connection to the personal and social life-world of people and that aims at restoring the harm resulting from the crime as completely as possible; (b) Restorative justice seeks to balance the needs of all involved. This inclusive approach entails bringing together the victim, the offender, the community and other actors in a common response where the "justice needs" of all stakeholders are addressed; (c) Parties in restorative justice are considered moral subjects able to participate in a process of dialogue and to encounter the other, on the condition that there is a mutually respectful environment created and an appropriate space and support provided for the encounter.

To deal with those principles we need to ask, with Aertsen (2018), "how does society respond to corporate violence, taking into account the specific nature of the effects, impact and meaning of this type of offence to the lifeworld of victims, their families and their communities?" One of the main problems that follows this question is how the criminal justice system ignores, or even intensifies and multiplies, the harms perpetrated by the powerful. At the same time, the penalties are usually applied to individuals, and the companies continue working afterwards, sometimes with the same behaviour that produced harm in the past (Aertsen, 2018, p. 240).

Although restorative justice seems to be potentially interesting in those cases, some limitations have also been addressed. One of the challenges is to deal with the traditional categories of "crimes", "victims", "criminals", "harm", and "community". Identifying environmental harm is complex because—as anticipated—most of the harms perpetrated are not consequences of behaviours classified in the juridical definition of *crime*. In the same sense, it is difficult to define who the *victims* are. Moreover, when dealing with harm to nature and to non-human animals, who would represent them in restorative justice practices? Aertsen (2018) observes that the people affected must be present in the restorative justice practices, as natural resources are mostly communal, for example "air, water and forests, and unowned property (such as light), public infrastructure, heritage, environmental meaning (sense and use of the environment by a community) and future generations". Here, "significance and subjective meaning are decisive. In the aftermath of the crime, meaning making will then be an important element of the recovery process, in which the experience and recognition of wrongfulness become key in the person's perception of justice" (Aertsen, 2018, p. 238).

Even to deal with human victims requires that there is a complexity about the causality nexus, for example in the long-term effects of pollution and exposure to toxic substances, as in the asbestos case. The strategies of denial in these cases have been successful for decades, also through the funding of biased research (Budó, 2021). And who are the *perpetrators*? Restorative justice in cases of corporate violence should also be seen as a way to prevent the company from perpetrating repeated harms, and not only to make an individual accountable. In the case of restorative justice, according to Braithwaite (2002) the more important motivational factor for the companies is to keep a reputation and a public image. The author proposes that *restorative shaming* could be an interesting possibility for offering a base for accountability in cases of corporate crimes and regulatory infractions. In this sense, the concepts of truth and memory are combined with the idea of exposing the image of the company, publicising the harm perpetrated and recognising the victims. At the same time, the role of the state is not so clear in many of those cases, but for sure it is also a stakeholder. For example, in the asbestos case, state collusion is clear in the avoiding of regulation. In the case of Belo Monte, the project was state-owned, and the state was also responsible for deciding where the hydroelectric dam would be constructed, without listening to the citizens who lived there.

All those difficulties raise the reflection that environmental restorative justice theory would benefit from the principles of transitional justice, mostly when it reinforces the complexity of the harms throughout time and space (Killean, 2022). The idea of a transgenerational justice, which deals with reparation, recognition, and memory, seems to adapt very well to the decolonial approach we propose here (Rodríguez, 2022).

This chapter had the objective of initiating a dialogue between the claims for justice brought from environmental victims' discourse and the principles of restorative justice. Restorative justice is a movement already well developed in many countries in the world, but it is also very different in each place where this label is used. The use of restorative practices in environmental harms and crimes is a new approach, which is being called "environmental restorative justice". When the green crime or harm is caused by a corporation, many complexities can be brought to the debate, as, for example: the problem of who will participate in the processes of dialogue both as the perpetrator (an individual or a company?) and the victim (only humans? the direct and/or indirect victims?); the problem with the possible use of a restorative circle as a strategy of greenwashing without repairing the harms to the victims; the problem with prevention, as companies are structurally stimulated to calculate environmental harms as collateral costs of production (Whyte, 2020). In this brief chapter we couldn't address all these complexities, which will be the focus of future research.

Going beyond the punitive logics for dealing with socio-environmental harms, searching for alternative criminal policies involving the need of acknowledgement of the victims, reparation and harm prevention seems crucial to confront the environmental destruction that is produced as an irreparable consequence of the process of capital accumulation.

References

Aertsen, I. (2018). Restorative justice for victims of corporate violence. In G. Forti, C. Mazzucato, A. Visconti, & S. Giavazzi (Eds.), *Victims and corporations: Legal challenges and empirical findings* (pp. 235–258). Wolters Kluwer Italia.

Allen, B. L. (2018). Strongly participatory science and knowledge justice in an environmentally contested region. *Science, Technology, & Human Values, 43*(6), 947–971.

Andrade, V. R. P. (2003). *A ilusão de segurança jurídica: do controle da violência à violência do controle penal.* Livraria do Advogado.

Auyero, J., & Swistun, D. A. (2009). *Flammable: Environmental suffering in an argentine shantytown.* Oxford University Press.

Baratta, A. (1985). Por una teoría materialista de la criminalidad y del control social. *Estudios Penales y Criminológicos* (Vol. XII).

Baratta, A. (2002). *Criminologia crítica e crítica do direito penal*. Revan.

Bisschop, L., & Vande Walle, G. (2013). Environmental victimisation and conflict resolution: A case study of e-Waste. In R. Walters, D. Westerhuis, & T. Wyatt (Eds.), *Emerging issues in green criminology: Exploring power, justice and harm* (pp. 34–56). Palgrave Macmillan.

Böhm, M. L. (2019). *The crime of maldevelopment: Economic deregulation and violence in the global south.* Routledge.

Bonilla-Silva, E. (2012). The invisible weight of whiteness: The racial grammar of everyday life in contemporary America. *Ethnic and Racial Studies, 35*(2), 173–194.

Borges, V. T. (2016). Carandiru: os usos da memória de um massacre. *Revista Tempo e Argumento, 8*(19), 04–33.

Braithwaite, J. (2002). *Restorative Justice and Responsive Regulation.* Oxford University Press.

Budó, M. D. (2019). "Um massacre silencioso que continua": um olhar criminológico sobre danos sociais causados pelo amianto. *Novos Estudos Jurídicos, 24*(2), 483–513.

Budó, M. D. (2021). Corporate crime and the use of science in the case of asbestos: Producing harm through discursive shields. *Journal of White Collar and Corporate Crime, 2*(2), 81–96.

Budó, M. D., & Falavigno, C. (2020). A tutela penal do meio ambiente: discussões criminológicas e dogmáticas. *Revista Brasileira De Ciências Criminais, 163*, 311–345.

Budó, M. N., & Pali, B. (2023). Restorative responses to harm caused by asbestos companies. *Journal of Victimology, 15*, 171–204. https://doi.org/10.12827/RVJV.15.06

Butler, J. (2011). Vida precária. Revista de Sociologia da UFSCar. São Carlos, Departamento e Programa de Pós-Graduação em Sociologia da UFSCar.

Carrington, K., Hogg, R., & Sozzo, M. (2016). Southern criminology. *The British Journal of Criminology, 56*(1), 1–20. https://doi.org/10.1093/bjc/azv083

Chateauraynaud, F., Debaz, J. (2022). Plurality of temporalities, complexity and contingency in repairing after dam failures in Minas Gerais. In S. Topçu, L. Centemeri, J. P. Burgess (Eds.), *Rethinking post-disaster recovery socio-anthropological perspectives on repairing environments*. Routledge.

Cohen, S. (1985). *Visions of social control*. Polity Press.

Cohen, S. (2001). *States of denial: Knowing about atrocities and suffering*. Polity Press.

Colognese, M. M. F. (2018). O caso Samarco: vitimização ambiental e dano social corporativo no cenário de Mariana—uma investigação empírica a partir da perspectiva das vítimas (parte I). *Revista Eletrônica Direito e Política, 13*(2), 956–988.

Connell, R. (2007). *Southern theory: The global dynamics of knowledge in social sciences*. Routledge.

Cruz, F. K. T., Giongo, C. R., Marques, G. S., Mendes, J. M. R. M. (2020). Isso é tirar a vida das pessoas: barragens e violações de direitos. *Psicologia & Sociedade*.

Davies, P., Francis, P., & Wyatt, T. (Eds.). (2014). *Invisible crimes and social harms*. Palgrave Macmillan.

Davies, P., Leighton, P., & Wyatt, T. (2021). *The Palgrave handbook of social harm*. Palgrave Macmillan.

Ebus, B., & Kuijpers, K. (2016). The state-corporate tandem cycling towards collision: State-corporate harm and the resource frontiers of Brazil and Colombia. In A. Brisman, N. South, & R. White (Eds.), *Environmental crime and social conflict* (pp. 125–152). Routledge.

Fainguelernt, M. B. (2020). Impactos da Usina Hidrelétrica de Belo Monte: uma análise da visão das populações ribeirinhas das reservas extrativistas da Terra do Meio. *Civitas-Revista De Ciências Sociais, 20*, 43–52.

Farias, E. (2014). "Herança" da ditadura militar no país ainda permanece na Amazônia. Amazônia Real, 31, mar. Disponível. https://amazoniareal.com.br/heranca-da-ditadura-militar-no-pais-ainda-permanece-na-amazonia/

Federici, S. (2009). *Caliban and the witch: Women, the body and primitive accumulation* (3rd ed.). Autonomedia.

França, K. A. (2022). *"Da lama ao caos": intersecções entre a criminologia verde do sul e a justiça de transição no testemunho dos atingidos e atingidas pelo dano social estatal-corporativo em brumadinho/MG*. Pontifícia Universidade Católica de Porto Alegre/RS.

Franko, K., & Goyes, D. R. (2023). *Victimhood, memory, and commercialism: Profiting from Pablo*. Oxford University Press.

Freitas, C. M., & da Silva, M. A. (2019). Acidentes de trabalho que se tornam desastres: os casos dos rompimentos em barragens de mineração no Brasil. *Revista Brasileira de Medicina do Trabalho, 17*(1), 21–29.

Furuya, S., Chimed-Ochir, O., Takahashi, K., David, A., & Takala, J. (2018). Global asbestos disaster. *International Journal of Environmental Research and Public Health, 15*(5), 1000. https://doi.org/10.3390/ijerph15051000

Gago, V. (2019). La potencia feminista: o el deseo de cambiarlo todo. Traficantes de Sueños.

Galtung, J. (1969). Violence, peace, and peace research. *Journal of Peace Research, 6*(3), 167–191. http://www.jstor.org/stable/422690

Galtung, J. (1990). Cultural violence. *Journal of Peace Research, 27*(3), 291–305. https://www.jstor.org/stable/423472

Ginzburg, J. (2001). Escritas da tortura. *Diálogos Latinoamericanos, 3*, 131–146.

Globus, R., Szasz, A., & Haluza-DeLay, R. (2014). *How the world's religions are responding to climate change: Social scientific investigations*. Routledge.

Gorostiza, S., & Armiero, M. (2021). *Repairing as struggle for narrative justice: The dam failure of Vega de Tera, Spain (1959–2019)*. Routledge.

Goyes, D. R. (2019). *Southern green criminology: A science to end ecological discrimination*. Emerald.

Goyes, D. R., & Franko, K. (2022). Profiting from Pablo: Victimhood and commercialism in a global society. *The British Journal of Criminology, 62*(3), 533–550.

Goyes, D. R., South, N., Ramos, D. T., Cuchimba, R. Ñ., Cuchimba, A., Baicué, P., & Abaibira, M. A. (2022). 'An incorporeal disease': COVID-19, social trauma and health injustice in four Colombian Indigenous communities. *The Sociological Review, 71*(1), 105–125.

Hall, M. (2013). *Victims of environmental harm: Rights, recognition and redress under national and international law*. Routledge.

Hall, M., & Varona, G. (2018). La victimología verde como espacio de encuentro para repensar la otredad más allá de la posesión. *Revista De Victimología, 7*, 107–128.

Halsey, M. (2004). Against "Green" Criminology. *British Journal of Criminology, 44*(6), 833–853.

Halsey, M. (2006). *Deleuze and environmental damage: Violence of the text*. Ashgate.

Haraway, D. (1988). Situated knowledges: The science question in feminism and the privilege of partial perspective. *Feminist Studies, 14*(3), 575–599. https://doi.org/10.2307/3178066

Harding, S. (1993). Rethinking standpoint epistemology: What is "Strong objectivity"? In L. Alcoff & E. Potter (Eds.), *Feminist epistemologies*. Routledge.

Hillyard, P., & Tombs, D. (2004). Introduction. In P. Hillyard, D. Tombs, & C. Pantazis (Eds.), *Beyond criminology: Taking harm seriously* (pp. 1–9). Pluto Press.

Killean, R. (2022). Environmental restorative justice in transitional settings. In B. Pali, F. Tepper, & M. Forsyth (Eds.), *The Palgrave handbook of environmental restorative justice* (pp. 247–274). Palgrave.

Krenak, A. (2019). *Ideias para adiar o fim do mundo*. Cia. das Letras.

Le Ven, M. M. (2014). Para que não se esqueça, para que nunca mais aconteça! 50 anos do golpe civil-militar: a Igreja e a universidade, Núcelo de Estudos Sociopolíticos.

Lynch, M., & Stretesky, P. (2003). The meaning of green: Contrasting criminological perspectives. *Theoretical Criminology, 7*(2), 217–238.

Mate, R. (2015). Memoria y justicia transicional. In J. G. S. Sousa Junior, J. C. M. Silva Filho, C. Paixão, L. G. D. Fonseca, T. T. D. Rampin (Eds.), *O Direito Achado na Rua vol. 7: Introdução crítica à justiça de transição na América Latina* (pp. 153–163). UnB.

Mbembe, A. (2003). *Necropolitics* (L. Meintjes, Trans.). *Public Culture, 15*(1), 11–40.

Natali, L. (2013). The contemporary horizon of green criminology. In A. Brisman & N. South (Eds.), *Routledge international handbook of green criminology*. Routledge.

Natali, L. (2016). *A visual approach for green criminology: Exploring the social perception of environmental harm*. Palgrave Macmillan.

Natali, L. (2021). Visual and sensory methodologies to explore environmental harm and victimization. In P. Davies, P. Leighton, & T. Wyatt (Eds.), *The Palgrave handbook of social harm* (pp. 139–165). Palgrave Macmillan.

Natali, L., & de Nardin Budó, M. (2018). A sensory and visual approach for comprehending environmental victimization by asbestos industry in Casale Monferrato. *European Journal of Criminology, 16*(6), 708–726.

Natali, L., & Hall, M. (2022). A green criminological approach to environmental victimization and reparation: A case for environmental restorative justice. In S. Topçu, L. Centemeri, J. P. Burgess (Eds.), *Rethinking post-disaster recovery: Socio-anthropological perspectives on repairing environments*. Routledge.

Natali, L., Suman, A., & De Nardin Budó, M. (2023). What about Environmental 'Victims'? Methodological Reflections for an Activist Criminology. In V. Canning, G. Martin and S. Tombs (Eds.), *International handbook of activist criminology* (pp. 215–231). Emerald.

Orth, G. M. N., Bourguignon, J. A., & Graf, P. M. (2020). O Sul também existe: intersecção entre o pensamento suleador e as práticas restaurativas no brasil. In G. N. M. Orth, P. M. Graf (Orgs.), *Sulear a justiça restaurativa* (pp. 19–43). Ponta Grossa, Brazil: Texto e Contexto.

Pali, B., & Aertsen, I. (2021). Inhabiting a vulnerable and wounded earth: Restoring response-ability. *International Journal of Restorative Justice, 4*(1), 3–16.

Pali, B., Tepper, F., & Forsyth, M. (Eds.). (2022). *The Palgrave handbook of environmental restorative justice*. Palgrave.

Pemberton, A., Mulder, E., & Aarten, P. G. M. (2019). Stories of injustice: Towards a narrative victimology. *European Journal of Criminology, 16*(4), 391–412.

Pemberton, A., Winkel, F. W., & Groenhuijsen, M. (2007). Taking victims seriously in restorative justice. *International Perspectives in Victimology, 3*(1), 4–14.

Piralian, H. (2000). *Genocidio y transmisión*. Fondo de Cultura.

Pires, A. (1999). Alguns obstáculos a uma mutação humanista do direito penal. *Sociologias, 1*(1), 64–95.

Pulido, L. (2018). Racism and the Anthropocene. In G. Mitman, R. Emmett, & M. Armiero (Eds.), *The remains of the Anthropocene* (pp. 116–128). University of Chicago Press.

Quijano, A. (2000). Coloniality of power and eurocentrism in Latin America. *International Sociology, 15*(2), 215–232. https://doi.org/10.1177/0268580900015002005

Reis, V. (2019). Nota Abrasco sobre a perversa tragédia da Vale em Brumadinho. Disponível em: https://www.abrasco.org.br/site/outras-noticias/notas-oficiais-abrasco/nota-abrasco-sobre-a-perversa-tragedia-da-vale-em-brumadinho/39675/

Rodríguez, I. (2022). Restor(y)ing the past to envision an 'other' future: A decolonial environmental restorative justice perspective. In B. Pali, F. Tepper, & M. Forsyth (Eds.), *The Palgrave handbook of environmental restorative justice* (pp. 531–562). Palgrave.

Rosa, M. F. (2016). *Os Atingidos de Belo Monte: experiências de sofrimento e agravos à saúde no contexto de um megaprojeto hidroelétrico na Amazônia brasileira.* Universidade de Coimbra.

Ruggiero, V., & South, N. (2013). Green criminology and crimes of the economy: Theory, research and praxis. *Critical Criminology, 21*(3), 359–373.

Saad-Diniz, E. (2019). Justicia restaurativa y desastres socioambientales en Brasil. *Revista De Derecho Penal y Criminología, 9*(10), 9–25.

Santos, B. S. (2016). *Epistemologies of the South.* Routledge.

Seligmann-Silva, M. (2008). Narrar o trauma: A questão dos testemunhos de catástrofes históricas. *Psicologia Clínica, 20*(1), 65–82.

Senado. (2019). CPI de Brumadinho. Relatório Final contemplando alterações aprovadas na última reunião da comissão. Disponível em: https://legis.senado.leg.br/comissoes/mnas?codcol=2246&tp=4

Silva Filho, J. C. M. (2012). Crimes do Estado e justiça de transição. In R. C. Gauer (org.), *Criminologia e sistemas jurídico-penais contemporâneos* (pp. 45–68). EDIPUCRS.

Silva Neto, N. (2021). Restorative Justice and Socioenvironmental conflicts in the Brazilian Amazon.

Silveira, A. M. (2018). Dano social estatal-corporativo e a vitimização ocasionada pela exposição ao amianto na cidade de Osasco-SP: um estudo criminológico a partir da representação das vítimas. Dissertação (Mestrado em Direito)—Curso de Pós-Graduação em Direito, Faculdade Meridional IMED.

Silveira, A. M., & Budó, M. D. (2021). Nuvem de Poeira: a experiência das vítimas ocupacionais e ambientais da indústria do amianto em Osasco—SP. *Revista Brasileira De Ciências Criminais, 182*, 229–260.

Sollund, R. (2022) 2. A vitimização através do tráfico e comércio de mulheres, crianças e espécies não humanas: crimes compreendidos sob uma perspectiva ecofeminista. In M. D. Budó, D. Goyes, L. Natali, R. Sollund, A. Brisman (Eds.), *Introdução à Criminologia Verde: Perspectivas Críticas, Decoloniais e do Sul* (pp. 289–308). Tirant.

South, N. (1998). A green field for criminology? A proposal for a perspective. *Theoretical Criminology, 2*(2), 211–234.

South, N. (2014). Green criminology: Reflections, connections, horizons. *International Journal for Crime, Justice and Social Democracy, 3*(2), 5–20.

South, N., Brisman, A., & Beirne, P. (2013). A guide to a green criminology. In N. South & A. Brisman (Eds.), *Routledge international handbook of green criminology* (pp. 27–42). Routledge.

Spivak, G. C. (1994). *'Can the subaltern speak?' Colonial discourse and post-colonial theory: A reader, edited by Patrick Williams and Laura Chrisman* (pp. 66–111). Columbia University Press.

Topçu, S., Centemeri, L., & Burgess, J. P. (Eds.). (2022). *Rethinking post-disaster recovery socio-anthropological perspectives on repairing environments*. Routledge.

Varona, G. (2020). Restorative pathways after mass environmental victimization: Walking in the landscapes of past ecocides. *Oñati Socio-Legal Series, 10*(3), 664–685.

Varona, G. (2022). Restorative justice for illegal harms against animals: A potential answer full of interrogations. In B. Pali, F. Tepper, & M. Forsyth (Eds.), *The Palgrave handbook of environmental restorative justice* (pp. 305–332). Palgrave.

Weinstock, A. M. (2017). A decade of social and environmental mobilization against mega-mining in Chubut, Argentinian Patagonia. In D. R. Goyes, H. Mol, A. Brisman, & N. South (Eds.), *Environmental crime in Latin America: The theft of nature and the poisoning of the land* (pp. 141–162). Palgrave.

Weis, V. V., Katz, M., & Martínez, S. (2022). Seletividade Penal Ambiental e Criminologia Verde do Sul: o caso de Andalgalá. In M. D. Budó, D. Goyes, L. Natali, R. Sollund, A. Brisman (Eds.), *Introdução à Criminologia Verde: Perspectivas Críticas, Decoloniais e do Sul* (pp. 75–90). Tirant.

White, R. (2008). *Crimes against nature: Environmental criminology and ecological justice*. Willan Publishing.

White, R. (2011). *Transnational environmental crime: Towards an eco-global criminology*. Routledge Publishing.

White, R. (2017). 'Reparative justice, environmental crime and penalties for the powerful', crime, law and social change. *An Interdisciplinary Journal, 67*(2), 117–132.

Whyte, D. (2020). *Ecocide: Kill the corporation before it kills us*. Manchester University Press.

Williams, C. (1996). An environmental victimology. *Social Science, 23*(1): 16–40. Reprinted in White, R. (Ed.) (2009). *Environmental crime: A reader* (pp. 200–222). Willan Publishing.

World Health Organization. (2010). *Environment and health risks: A review of the influence and effects of social inequalities*. WHO.

Wyatt, T. (2022). *Wildlife trafficking: A deconstruction of the crime, victims and offenders* (2nd ed.). Palgrave.

10

Pop Culture as Environmental Education in Japan: The Case of Hayao Miyazaki's *Kaze-no-tani-no-Naushika*

Orika Komatsubara

Introduction

Despite the knowledge that humans have regarding environmentally sound behaviour, we seem to continue acting foolishly. For example, a survey conducted in Belgium in 2022 showed that whilst young people are aware of the environmental harm and greenwashing committed by clothing companies, they continue buying clothes from fast fashion companies (Komatsubara et al., 2022). While criticising people's cognitive dissonance is one way of confronting environmental harm, I suggest a more positive alternative: in-depth environmental education focused not on imparting knowledge as to environmental harms but instead on stimulating people's emotions and senses, working directly on their hearts and minds.

O. Komatsubara (✉)
Osaka Metropolitan University, Osaka, Japan
e-mail: kmtbrllac@gmail.com

D. R. Goyes (ed.), *Green Crime in the Global South*, Palgrave Studies in Green Criminology, https://doi.org/10.1007/978-3-031-27754-2_10

Art can profoundly impact people's minds and stir their hearts. While traditional green criminology has focused primarily on the *objective realities* of environmental destruction, policy analysis, and the visible actions of social movements, Brisman and South (2017) suggest a green *cultural* criminology that focuses on the dramaturgy of environmental crime, including representations in science fiction, reality television, documentary films, and mockumentaries. For them, fictional stories can potentially convey moral messages about the relationship between humans and the environment (Brisman & South, 2017).

Intending to further develop a Southern green cultural criminology (see Goyes et al., 2021) that plays on the strengths of popular knowledge, I will argue that art has the potential to convey the experiences of victims of environmental degradation effectively. I focus on Hayao Miyazaki's full-length manga (graphic novel), *Kaze no Tani no Naushika* (*KTN*, 1982–1995). Though manga and animation are a dominant form of pop culture in Japan and have a growing international presence, there are few examples of research involving them in green criminology. However, pop culture, as its name indicates, is accessible to a wide audience. If pop literature can be used to convey the realities and experiences of environmental harm, we can share the seriousness of this issue with a much larger public audience than scholarly work does.

In this chapter, first, I provide the role of manga and anime in Japanese pop culture, and explain why *KTN* is an appropriate educational tool to increase the understanding of environmental harm. Second, I elaborate on how *KTN* stimulates the reader's sensibility through non-verbal expressions and activates the mind's eye via illustrations. I further explain why Miyazaki's art encourages the reader to share the protagonist's struggles in the face of environmental destruction. Third, I interpret the relationship between *KTN*'s protagonist and non-human beings by highlighting the element of intimacy between them (the protagonist suffers from an ethical conflict about prioritising which human or non-human interests). Finally, I explain how this piece of popular art—like many others—can serve as a valuable resource for environmental education.

Manga and Animation as the Dominant Form of Pop Culture in Japan

Anime and manga are important export products in Japan. In contrast to the relatively low presence of Japanese music and live-action films in the international entertainment industry, Japanese anime and manga generated USD51.4 billion in sales in 2016, and the figure is expected to increase (Ministry of Economy, Trade and Industry of Japan, 2017). The Japanese government launched its 'Cool Japan' policy in 2010 to incentivise the international promotion of anime and manga.[1] In addition to expanding the international circulation of manga and anime, international researchers contributed to the global interest in those genres when they turned their academic gaze towards them in the late 1990s (Yamada, 2017). Contemporary Japanese manga and anime depict everyday life and communication in minute detail; therefore, they can challenge stereotypical images of Japanese culture, such as geishas, sumo wrestlers, and samurai warriors. While Walt Disney and other international actors have significantly influenced Japanese anime and manga, these forms have in turn been instrumental in identifying characteristics of non-mainstream Japanese culture and society, as well as in criticising the traditional stereotypical view of 'the Japanese' (Yamada, 2017).

Hayao Miyazaki, a well-known animation director and co-founder at Studio Ghibli[2]—an acclaimed Japanese animation film studio—has conducted social criticism through his works, mainly regarding environmental justice. Miyazaki explores utopian worlds based on Japanese

---

[1] Within Japan, the 'Cool Japan' strategy has not been welcomed by the majority. Manga and anime have developed at the initiative of the private sector rather than the state (Mihara, 2014). Notably, the Cool Japan Organisation, funded by the Japanese Government, has failed, as it had an accumulated deficit of JPY30.9 billion (USD2.2 billion) in 2022 and is being considered for fusion with other organisations to rebuild it (Yomiuri Shinbun, 2022). In addition, the Japanese government has a history of exploiting manga and anime for political purposes. During the Second World War, those art expressions were used to elevate nationalism, colonial rule, and overseas propaganda. The 'Cool Japan' strategy is an extension of that trajectory (Otsuka, 2022).

[2] Studio Ghibli is also affiliated with another important animation director, Isao Takahata. Although Takahata is not discussed in this chapter, his film *Heisei Tanuki Gassen Pompoko* (*Pom Poko*, 1994) is an interesting story of a resistance movement against humans by raccoon dogs that have lost their homes owing to housing development. The film shares the perspective of non-speciesist environmental justice.

animism, wherein human beings and nature co-exist in harmony (Akasaka, 2019; Koyama, 2009; Napier, 2019; Nomura, 2018; Sugita, 2014; Yoneyama, 2019). In *Tonari no Totoro* (*My Neighbour Totoro*, 1988), Miyazaki depicted interactions between the children in a rural village and the gods and spirits that live in the nearby forest. *Mononoke-hime* (*Princess Mononoke*, 1997) explores the same type of co-existence by presenting a story set in medieval Japan depicting a battle between human beings seeking technological development and gods living in a forest. *Sen to Chihiro no Kamikakushi* (*Spirited Away*, 2001) tells of the adventure of a young girl who wanders into another world and rescues a river god contaminated by human beings. Children love Miyazaki's animated films, which have been repeatedly broadcast in Japan and elsewhere. His work is often used as a resource for environmental education (Tanaka et al., 2012). In November 2022, Ghibli Park opened in Japan, allowing children to experience the world of Miyazaki's animation. Ghibli Park was designed not as a large attraction but as a facility where visitors can walk through a forest and reconsider their relationship with nature. Miyazaki's works have thus served as a bridge between Japanese pop culture and environmental education.

One of the features of Miyazaki's animism is that it is born out of his own experience, not from theories of environmental philosophy. Miyazaki was born in 1941, during the Second World War. After the war, amid Japan's rapid economic development, he witnessed people losing interest in nature as they pursued a new, affluent society. In learning about the plant ecology of the shiny-leaved forests spread across East Asia, Miyazaki came to understand the symbiotic relationship between human beings and nature (Miyazaki, 1996). Nevertheless, he kept his belief in the unexplainable sensation of 'something being there' when walking in the forest—a feeling that lies at the core of Japanese animism (Miyazaki, 1996). Respect for nature, born out of a sense of wonder, is part of Miyazaki's animistic sensibility. Through his art, he conveys that indescribable *something* of the gods and spirits inhabiting nature, which cannot be scientifically proven or even verbalised. His work is characterised by eliciting animistic sensibilities, by emotionally stimulating the audience via moving pictures and sound, rather than by explaining issues logically to seek justice. That is, he conveys the world of animism to the audience through non-verbal artistic expression.

Miyazaki's works are internationally recognised and highly acclaimed. Japanese scholar Susan Napier held a year-long seminar on Miyazaki's work at Tufts University in the USA, and reported that young US students exposed to his work were sensitively stimulated (Napier, 2019). Internationally, the most commercially successful of his works is *Spirited Away*, which was released in Hong Kong, France, and South Korea in 2001 and became widely popular. In 2002, the film won the Golden Bear at the Berlin International Film Festival. It was the first time in history that a full-length animated film had won at all three major film festivals (Cannes, Venice, and Berlin). Furthermore, in 2002, the film was released in North America, where it won the Critics' Choice Movie Awards and was named one of the Top 10 Films in 2002 by *The New York Times*, *Newsweek*, *Time*, and *The Wall Street Journal* (Kano, 2006). Finally, the work won the Academy Award for Best Animated Feature.

Miyazaki is undoubtedly one of the most internationally renowned animation directors. While outstanding as a director, the medium of animation also has its limitations. Animated films are made by linking illustrations,[3] playing music, and voicing characters to create moving images. Animation requires co-production, employing highly skilled staff. Additionally, it must be marketed to cinemas to attract an audience. Thus, production costs are enormous. Animated films must make a profit, making commercial success their goal. Although Miyazaki has had a strong track record of commercial success, he decided to draw the manga *KTN* alongside his work as an animation director. Manga can be produced by a single person using manuscript paper and drawing materials.[4] *KTN* was published in *Animage*, whose primary readers are adults. It has also been published in English, but is less known internationally compared to his animated works. However, looking at this manga

[3] Contemporary animation generally uses computer graphics, but the animation of *Kaze no Tani no Naushika* was drawn by hand, one illustration at a time, by humans—this is known as cel-animation.

[4] Contemporary manga is increasingly executed by relying on a specialist division of labour, though the manga *Kaze no tani no Naushika* was largely executed as an individual production.

provides in-depth insight into Miyazaki's environmental thought, which I examine next.

Kaze no Tani no Naushika and Environmental–Ethical Conflict

KTN is set in a post-apocalyptic world, following a war. The air and the land are contaminated by chemical weapons, and people must wear masks. The main character, a young girl named Naushika, is the daughter of a small local community leader. She loves nature and wants to protect it. In the main storyline, she tryies to negotiate with imperial politicians to stop the war and lead the people towards a symbiotic relationship with nature. *KTN* has two versions: a manga version and an animated version. The manga version (1982–1995) of *KTN* is over 1000 pages long, with all its illustrations drawn by Miyazaki himself. Comparing the manga and animated versions shows that Miyazaki drew the manga differently than it was animated (Miyazaki, 1996). The animated version of *KTN* (Miyazaki, 1984) was directed by Miyazaki and released. Following the lead of Studio Ghibli producers, Miyazaki deconstructed the manga version of Naushika's initial episodes and reconstructed them into a 116-minute animated film. The animated version of Naushika has a vivacious, forward-looking, and sparkling-eyed girl as the protagonist. She is angry at the destruction of nature and saddened by its demise. She sacrifices herself and dies trying to save both humanity and nature—but supernatural powers resurrect her. Impressed by Naushika's actions, other people follow her lead. She becomes a kind of environmentalist leader capable of stopping the human destruction of nature and leading the world towards interspecies peace, an icon of environmental change. Therefore, unsurprisingly, the World Wildlife Fund endorsed the film.

In contrast, in the manga version, Naushika is morally ragged and filled with doubt. Human destruction of nature represents a weight she seems incapable of bearing, and leads her to wish to give up her humanity and become one with nature. However, she is the leader of her small local community and is respected by all. Given her responsibility to her people, the dilemma of whether to prioritise the protection of nature or the human community splits and depresses her (*KTN*, 5). One manga technique is to place spoken words in speech bubbles and internal monologues in the margins. In the manga version of *KTN*, monologues appear more frequently; Naushika is less talkative with her peers and friends but has long introspective reflections (*KTN*, 6). In the end, Naushika encourages her people by telling them that the polluted world will be purified and the human race will survive. However, her internal monologue reveals a different picture: 'I told a lie to the people. I will continue to do so' (*KTN*, 7, p. 171). She knows that people's bodies, including her own, have adapted to the contaminated earth and sky, unable now to survive in a purified world; nevertheless, she shows her people a false future to give them hope whilst prioritising the restoration of nature. She does not confide in anyone her thoughts about the ideal relationship between humans and nature—she reflects and acts alone (*KTN*, 7). Her only confidant is the manga reader, who is deeply aware of her ethical conflicts.

Miyazaki has said that the animated film of *KTN* was made for entertainment (Miyazaki, 1996). He further stated that in the manga version, the story developed beyond his control: He did not intend to produce an eco-philosophical work, but when he got to the heart of the relationship between nature and human beings, he realised there were no easy answers. As he dug deeper, he was confronted with complex questions—like the ancient monks who trained in the mountains who thought: 'What is the universe?' 'What is life?' Whilst Miyazaki avoided quoting religious leaders, not wanting to risk looking like an imitator or follower (Miyazaki, 1996), he was inspired by them and began to 'feel that there is a vast amount of experience and diverse elements hidden in the words of seemingly simple teachings written in a little introductory book on Buddhism', for example (Miyazaki, 1996, p. 527). So, he focused on the *lack of words*.

> It was exactly like Naushika herself was without words [when she was in that state]. I did not want to express her thoughts in words ... The moment I thought, 'Is this what she is thinking?' and put it into words, it seemed different. (Miyazaki, 1996, p. 527)

Therefore, in the manga version, Miyazaki focuses on expressing the state of 'wordlessness'. As discussed in the previous section, Miyazaki's animated films were appealing in their attempt to express the world of animism in a non-verbal way. In the manga version, it is not only the gods and spirits that are non-verbally represented but also Naushika's introspection and conflicts regarding her ideal relationship between humans and nature. Readers of the manga version witness Naushika staring blankly at the destruction of nature, with a frozen expression: no argument is outlined about the right thing to do in response to Naushika's conflict. Readers share the weight of her speechlessness. In the following sections, I will analyse specific scenes depicting her ethical conflict.

Ethical Dilemmas in Human–Nature Interactions

To analyse the ethical conflict experienced by Naushika in the manga version of *KTN*, I focus on the relationship between Naushika and the *Ohmu* (also called 'kings of bugs', fictional non-human beings). The *Ohmu* are insect-like creatures larger than humans, with hard shells and an exoskeleton, and who live in groups. They have several blue eyes on their backs, which turn red when they feel angry or threatened. Although Naushika loves the *Ohmu*, most people fear them because they often attack human towns and villages. Naushika, as the leader of a small rural community, has the responsibility to ensure a safe life for the people there. She is therefore faced with an ethical conflict between the interests of humans and the protection of the *Ohmu*. (A similar conflict is often experienced by people in the real world, and amounts to social justice versus species justice [Sandvik, 2022]). Below, I focus on two related motifs and describe their relationship in the story.

The first motif is the boundary between human and non-human beings. At a very young age, Naushika made friends with a little *Ohmu*. However, her parents attempted to kill the *Ohmu*, as they considered her to be mentally controlled by it. Naushika ran away from her parents and cuddled the young *Ohmu* while shrieking, 'Don't kill it'; but as a child, she could not defend it. Finally, it was killed by the other humans; her father said, 'bugs and people cannot live in the same world' (*KTN*, 1, p. 130). The scene generated trauma in Naushika and taught her about the strict barriers most people establish between humans and non-human beings. Her love for the *Ohmu* was from then onward a secret. At least, she tried to keep it a secret, but other adults discovered her love for the *Ohmu*, and regarded it as an extraordinary capacity. Simultaneously, 'that capacity does not help win wars at all. Rather, it is a power to heal wounds' (*KTN*, 1, p. 80). The adults in her community are concerned whether Naushika's care capacity during wartime is desirable. She is, after all, the community leader.

Naushika gets upset when she realises that the *Ohmu* are being exploited as a weapon of war, and resigns her role as leader to rescue the bugs. Seeking to elicit anger in adult *Ohmu* herds and induce them to attack (human) enemy troops, humans capture and injure a young *Ohmu*. Naushika then says, 'What a terrible thing.... It is unbelievable that the heart of bugs [*Ohmu*] trying to protect their family and friends is used [as a weapon] in war' (*KTN*, 2, p. 40). She fights an army of humans to save the baby *Ohmu*. However, its wounds are too deep. Naushika considers killing it to relieve the pain; she tries to shoot it with her gun, but gives up and says: 'I can't shoot you. No matter how much pain you are in, you are alive. You are trying to live' (p. 64). She hugs it, even though it is covered in blood, and vows that if it dies, she will stay by its side. The bug, however, wants to rejoin its herd family and friends, and tries to run again—risking falling into a dangerous acid lake in front of them. Naushika tries to stop it with a hug. In her efforts, Naushika injures her leg with the acid and screams in pain, her face contorted. As soon as the *Ohmu* looks at her face, it immediately tries to help and heal her wounds. She murmurs in her mind: '[*Ohmu*, you're] a sweet kid' (p. 66).

Finally, with the help of Naushika, the young *Ohmu* joins the herd. At that moment, each adult *Ohmu* extends its tremendous number of tentacles towards Naushika. They wrap around her body and lift her into the air. Their tentacles—which have the power to heal wounds—heal her leg. Countless tentacles glow golden and wavered around her, looking like a wheat field at harvest time. She tears up and looks around at the tentacles, grateful for their healing powers and calling out, 'Thank you, thank you, everyone' (*KTN* 2, p. 78). One girl who observed Naushika perceived that 'She [Naushika] was walking through a meadow at twilight' (p. 79). The *Ohmu* herd fully accepts Naushika, and her love is transmitted to them. In *KTN*, the emergence of a caring and emotional relationship between humans and non-human beings is illustrated through the interaction between Naushika and the *Ohmu*. Thus, the story describes how she crosses the boundary between human and non-human beings.

The second motif is the conflict between environmentalism and anthropocentrism. Later in the story, Naushika notices that trees grow from *Ohmu*'s bodies (*KTN*, 5). Trees make up the forest, absorb the toxins of human chemical weapons, and purify the earth. Then, she realises that the *Ohmu* had begun to move towards death in the south in their effort to clean up the world. She weeps:

> You [*Ohmu*] knew all along that this would happen [that human destruction of nature would have to be atoned for by your sacrifices]. There is no way to stop the stupid behaviour of humans. I understand that it is already too late for everything. But [I would ask you] not to leave me alone. (pp. 138–139)

Naushika suffers from feeling responsible, as a human being, for the mass death of the *Ohmu*. Then, she is called by a skeletal monster. The monster is called *Kyomu* (emptiness) and symbolises nihilism. *Kyomu* justifies the human-induced sacrifice of *Ohmu*.

> No need for you to be sad about the *Ohmu*'s death. Everyone will soon be free from suffering. They are fulfilling their divine role .... *Ohmu* was created to cleanse this planet of human pollution .... Look at your own feet. There are only dead people [under your soles]. God will not allow any more stupid people to pollute the earth. (pp. 140–141)

Kyomu's suggestion is to accept the doom of the world. He tempts Naushika saying that if the entire planet perishes due to the human destruction of nature, all beings will be freed from suffering through death. However, she refuses to do so and say 'No' against the justification of the human destruction of nature and fights in her inner world, even when *Kyomu* tries to persuade her. *Kyomu* plays Naushika's guilty conscience, accusing her of having survived the war while killing people.

> You have always pretended to be an innocent child, but it has been exposed as a lie; *Ohmu* will not tolerate you anymore. You are just one of the stupid and filthy humans. You are a human adult. You are a bloody woman of a cursed species. Go gallivant with the dead.

Although Naushika acknowledges that she, too, is a human who has survived at the expense of others, she resists *Kyomu*'s words and does not accept any justification for the world's demise at the hands of humans. She says:

> I do not need *Kyomu* to tell me that I understand that we are a cursed species. We are the ugliest of creatures, who only harm, rob, pollute and burn the earth. Bugs are much more beautiful than we are. What good is it to ask forgiveness for them now, when the *Ohmu* have made a long journey [to the south to die], blinded by mycelium, to heal the wounds caused by humans. That journey will soon come to an end. (p. 143)

Naushika embraces *Ohmu* and strives to get closer to them and get to know them. She learns that they do not think of taking revenge on the humans responsible for their sacrifice, but die for the purification of the world. As they die, new plants sprout from their bodies and grow into trees, and she hears the sound of the roots of the trees growing and breaking through the *Ohmu*'s flesh. The sound signifies the death of *Ohmu* and the birth of new life. Overwhelmed by the rapid death and rebirth cycle, she murmurs over their backs, 'I will go with you [*Ohmu*]' (p. 152). She chooses to die with the *Ohmu*; however, the *Ohmu* again extend their tentacles towards her, enveloping her and placing her inside their own body for protection, thereby saving her life.

Naushika continues to sleep inside *Ohmu*'s body and wanders through her mind's world. To save her, a man who calls himself *mori no hito* (the forest human) comes to her mental world with psychic powers. *Mori no hito* meets with Naushika and guides her on a spiritual journey to a secret forest (*KTN*, p. 6). Passing through the forest's depths, she discovers beautiful clear blue skies and uncontaminated plants. The landscape is beautiful, with grass fluttering and birds circling in the air. Swans are swimming in the pond, but at the first sign of humans, they rush off in panic. This was the natural landscape before the human-driven apocalypse occurred. Thus, Naushika finds hope, knowing that parts of nature are being revived. She is in awe of the power of nature. However, she immediately decides to leave there:

> I will return to my [polluted] world. We [humans] must not contaminate this world. How nice it would be if I could live here with everyone [in my local community] without worrying about the polluted land and the poison. But if the humans now know [this forest], we will begin to think that we are the masters of the world again. We will quickly invade this new-born land and do it all over again. If a thousand years or more had passed, and this world had become bigger and stronger, and we had become a bit more clever without being extinct, then we would come here then. (pp. 93–94)

Afterwards, Naushika awakens from sleep in the body of *Ohmu* and returns from the mind world to the real world. *Mori no hito* is standing in front of her, and he invites her to come to their community. He and his friends live away from the majority of human society and abide in harmony with nature in the forest. He feels Naushika has the same ethos as them and asks her to 'live with me' (p. 97). In other words, he suggests she choose an environmentalist lifestyle, leaving her local community. She looks surprised for a moment, hesitates, then bursts into tears and replies, 'Thank you, I'm very happy.' However, she looks away and continues her words, 'But, you are in the lifestream, while I am involved with individual life' (p. 97). Finally, she says, 'I love the humans in this [contaminated] world too much. I will live in a human, polluted, doomed world' (p. 98). She smiles slightly, but vaguely and unclearly. She appears both happy and unhappy about her decision. *Mori no hito* says, 'Indeed, it may be severe to take you away [from the contaminated world]. Too many beings [including non-human beings] love you' (p. 98).

In the next scene, Naushika's admirers come to her in a rush. When they see her, they shed tears and are happy to see her again. Elder members of the local community also come and embrace her. She chooses to take responsibility as a local community leader and prioritises her life with them. In *KTN*, she is constantly divided between environmentalism and anthropocentrism; her process of ethical conflict is described, until she ultimately chooses to live among the humans but to try to advance environmentalism.

Discussion

Hayao Miyazaki's *KTN* is useful for teaching environmental ethics in three senses. First, manga readers may experience an emotional transformation through the story. By immersing themselves in the story and witnessing Naushika's actions, readers can generate an emotional relationship with the *Ohmu*, as Naushika does. Noteworthily, readers do not start with a positive attitude towards the *Ohmu*, but rather develop one while reading the manga.

According to Miyazaki, the *Ohmu* are designed to symbolise creatures in ecosystems that humans are destroying. He assumed that bugs are difficult for humans to read emotionally and are not easy to feel compassion for (Miyazaki, 1996). In Japan, for example, there are dog and cat protection groups, while insecticides for slaughtering cockroaches can be found in supermarkets. If the *Ohmu* had the appearance of an animal like a dog or a cat, which are often our companions, and if it had been easier to read their emotional expressions, the reader's impression of the story would have been entirely different. Therefore, Miyazaki used their appearance as a gimmick, so that before reading the manga, the reader feels fear and hatred towards the *Ohmu*.

Thus, Miyazaki could position Naushika as an extremely rare person, who loves the *Ohmu*, which most people hate. Miyazaki modelled her on the *mushi mezuru hime gimi* princess who loves bugs in the *Tsutsumi Chunagon Monogatari*, a collection of short stories originating in thirteenth-century Japan (Miyazaki, 1996). The princess criticised people for hating the caterpillar while glorifying beautiful bugs like butterflies. The humans around her did not understand her ideas. Miyazaki added that 'She [Naushika] is a strange girl. She regards bug life and human life as equal' (p. 473). Because she is the protagonist, *KTN* readers follow the story from her side and through her eyes. Consequently, the reader shifts from disliking the *Ohmu* as most (including the people in Nushika's community) do, to loving them. In other words, by reading the manga, readers can have a vicarious experience of the perspectives and feelings of others regarding non-human beings. In this way, environmental education materials using pop culture have the potential to enable us to be more malleable in our views of non-human beings.

Second, environmental education using art differs from solution-oriented environmental ethics. Often, environmental justice and social justice are discussed dichotomously in environmental ethics. Naushika's ethical conflict between protecting the *Ohmu* and taking responsibility as a local community leader overlaps with the dichotomy between the two justices. Whilst she appears to finally make an ethical decision based on social justice, the key is to consider the internal process by which she reaches it. The sequences in the story where she experiences a spiritual journey guided by *mori no hito* and is exposed to the sublime depiction of nature are slow-paced and use a large number of pages with beautiful illustrations. Although she is attracted to the environmentalist lifestyle, she does not choose to become entirely an environmentalist. It is not lack of knowledge or community peer pressure that hinders her, but a deep ethical conflict which might resonate with many readers. Therefore, *KTN* readers are not instructed to go for either environmental or social justice, but incited to experience the pain of having to make a decision between them, thereby inviting deep reflection.

Through Naushika's interaction with several characters in *KTN*, readers learn different ideas about nature. The multiplicity of voices about the environment in the manga resemble the status of world affairs. In the face of cases of environmental harm in the real world, people have conflicting and sometimes intense clashes of opinion based on their interests and personal feelings. Meanwhile, fictional works allow readers to consider imaginary characters and environmental harm from a distance. Therefore, artworks such as *KTN* provide a medium for people to consider environmental harm from multiple perspectives.

Third, pop culture reaches a broad audience. Ecocritical studies have examined cases of outstanding literary works depicting environmental harm. Highly acclaimed artworks in the literary world can be valuable as material for environmental education. An example is the literary works of Michiko Ishimure on Minamata disease, a case of environmental pollution that has caused severe harm to humans and nature in Japan. Ishimure documented that a big company, prioritising economic profit over preventing the loss of human and animal life, had discharged polluted wastewater into the sea. In addition, Ishimure's works stimulated the emotions of her readers by describing the suffering and sorrow of the local people and their daily lives. Furthermore, her literature expresses the animistic world of *yokai* (spirits) and gods in the local culture (Komatsubara, 2021a, 2021b). Although her works brilliantly depict the actual lives and lifeworld of the Minamata disease victims, they are challenging for the reader to understand, with many abstract expressions and depictions of narratives in the dialects of Minamata disease victims. Her beautiful, fantastic descriptions stimulate the reader's imagination; however, they require the reader to have a literary background. Thus, despite her work's high value and quality, her readers are limited to the intellectual class and book lovers.

In contrast, 17 million copies of *KTN* have been sold as of 2020 (Bunka Tsushin, 2021). Miyazaki has said that *KTN* was inspired by the pollution in Minamata (Miyazaki, 1996). Though *KTN* is based on a real case of environmental destruction, many readers enjoy *KTN* simply for the entertainment. Unlike traditional literary works, manga is illustrated. Readers who cannot comprehend abstract texts or imagine concrete landscapes from metaphors can still find their senses influenced by the power of the illustrations. Further, it is easier for readers to empathise with a character's emotions if, as in manga, these emotions are drawn with simple lines. Thus, readers are drawn in by the manga and exposed to the theme of environmental destruction. Pop culture, such as *KTN*, is an accessible entry point for learning about environmental harm.

Conclusion

Hayao Miyazaki's *KTN* provides educational material to help us understand the realities and struggles of victims of environmental harm. The products of pop culture, such as manga, stimulate people's emotions and hold the potential to teach readers about environmental degradation positively and enjoyably. Art can be a tool for learning about other people's worldviews and can provide alternative introductions to environmental destruction, beyond solution-oriented content concerning the environmental crisis.

A limitation of this study is that it focuses on the positive efficacy of fictional works in environmental ethics, and does not consider their shortcomings. Two new films on Minamata disease were released in 2021 in Japan. One is a fictional work, *Minamata*, directed by Andrew Levitas (2020) and starring Johnny Depp as the American photographer Eugene Smith in the lead role. This dramatic film explores the causes of Minamata disease. However, it does not depict reality accurately; while Smith lived in Minamata, took important photographs of the victims, and reported them to the world, several historical facts were altered in the film to portray Smith as a hero. The other is the documentary film *Minamata Mandala*, filmed by Kazuo Hara (2021). The film describes the court battles and daily lives of the victims and supporters living in Minamata today. The director spent more than 15 years filming and completing this 372-minute epic. Critic Takayuki Kan (2021) compared the two films and asked whether it is facts or emotions that matter more to people. While documentaries recording the facts of real-life victims and their supporters are important, not many ordinary people would be willing to watch a 372-minute film. However, the attitude that prefers to indulge in entertainment films for emotional stimulation and disregards historical facts is a severe problem for environmental ethics. My future research agenda will include an examination of the dangers of people escaping real-world environmental harm by immersing themselves in artworks.

References

Akasaka, N. (2019). *Naushika ko: kaze no tani no mokushiroku*. Iwanami Shoten.

Brisman, A., & South, N. (2017). *Green cultural criminology: Constructions of environmental harm, consumerism, and resistance to ecocide*. Routledge.

Bunka Tsuhin (2021, July 11). Tokuma Shoten Komikkusu *Kaze no Tani no Naushika* ga Ruikei 1700 Manbu Toppa. (online). https://www.bunkanews.jp/article/226312/

Goyes, D. R., Abaibira, M. A., Baicué, P., Cuchimba, A., Ñeñetofe, D. T. R., Sollund, R., South, N., & Wyatt, T. (2021). Southern green cultural criminology and environmental crime prevention: Representations of nature within four Colombian indigenous communities. *Critical Criminology, 29*(3), 469–485. https://doi.org/10.1007/s10612-021-09582-0

Hara, K. (2021). *Minamata Mandara*. Shisso Prodakushon.

Kan, T. (2021). Jijitsu ka Kyoko ka ga Towareru. *Eiga Geijutsu, 477*, 24–25.

Kano, S. (2006). *Miyazaki Hayao Zensho*. Firumu Aatosha.

Komatsubara, O. (2021a). 'Kitsune ni Damasareru Chikara' o Torimodosu: Minamatabyo o Toshita Kankyo Kyoiku no Kanosei. *Gendai Seimei Tetsugaku Kenkyu, 10*, 96–118.

Komatsubara, O. (2021b). Imagining a community that includes non-human beings: The 1990s Moyainaoshi movement in minamata, Japan. *International Journal of Restorative Justice, 4*(1), 123–140.

Komatasubara, O., Pereira, A., & Sechidou, K. (2022). *Green criminology and environmental harm: In the case of fast fashion industry*. [Paper presentation] European Society of Criminology, 22nd Annual Conference, University of Malaga, Spain.

Koyama, M. (2009). *Miyazaki Hayao Mangaron: Kaze no Tani no Naushika Seidoku*. Gendai Shokan.

Levitas, A. (2020). *Minamata*. Long Ride and Albatross Films.

Mihara, R. (2014). *Kuru Japan ha Naze Kirawarerunoka: 'Nekyo' to 'Reisho' o Koete*. Chuo Koronsha.

Ministry of Economy, Trade and Industry of Japan. (2017). Heisei 29 Nendo Chiteki Zaisanken Waakingu Guruupu Tou Shingai Taisaku Kyoka Jigyo ni okeru Kontentsu Bunya no Kaigai Shijo Kibo Chosa. https://www.meti.go.jp/policy/mono_info_service/contents/downloadfiles/report/2017_kaigaishijochosa.pdf

Miyazaki, H. (1983–1995). *Kaze no Tani no Naushika*, 1–7. Tokuma Shoten. English Trans. (2012). *Nausicaä of the Valley of the Wind*. Viz Media.

Miyazaki, H. (1984). *Kaze no Tani no Naushika* (*Nausicaä of the Valley of the Wind*). Toppu Kruahuto.

Miyazaki, H. (1988). *Tonari no Totoro* (*My Neighbor Totoro*). Studio Ghibli.

Miyazaki, H. (1996). *Shuppatsuten: 1979–1996*. Tokuma Shoten.

Miyazaki, H. (1997). *Mononoke hime* (*Princess Mononoke*). Studio Ghibli.

Miyazaki, H. (2001). *Sen to Chihiro no Kamikakushi* (*Spirited Away*). Studio Ghibli.

Napier, S. (2019). *Miyazaki warudo: Miyazaki Hayao no hikari to yami*. Hayakawa Shobo. English version (2018): *Miyazaki World: a life in art*. Yale University Press.

Nomura, K. (2018). *Shinban Miyazaki Hayao no Chihei: Naushika kara Mononoke Hime made*. Shintensha.

Otsuka, E. (2022). *Daitoa Kyoeiken no kuru Japan: 'kyodo' suru bunka kosaku*. Shueisha.

Sandvik, K. B. (2022). The Ukrainian refugee crisis: Unpacking the politics of pet exceptionalism. *International Migration*. https://onlinelibrary.wiley.com/doi/10.1111/imig.13100

Shinbun, Y. (2022, June 21). Kuru Japan Kiko no Ruisekiakaji 309 Okuen, Kaizen Mirarenai Baai ha Tohaigo o: Zaimusho ga Teigen. (online) https://www.yomiuri.co.jp/economy/20220620-OYT1T50172/

Sugita, S. (2014). *Miyazaki Hayao ron: Kamigami to kodomotachi no monogatari*. NHK Shuppan.

Takahata, I. (1994). *Heisei Tanuki Gassen Ponpoko* (*Pom Poko*). Studio Ghibli.

Tanaka, M, Totsu, M., & Omori, A. (2012). *Webu Ankeeto o Riyoshita Kankyo Kyoiku Koka Sokutei no Teian*. [Paper presentation] Joho Shori Gakkai, 74th Zenkoku Taikai, Nagoya Institute of Technology, Japan.

Yamada, S. (2017). Manga Anime de Kenkyusuru to Iukoto. In Yamada (Ed.), *Manga Anime de Ronbun Repooto o Kaku*. Mineruba Shobo.

Yoneyama, S. (2019). *Animism in contemporary Japan: Voices for the Anthropocene from post-Fukushima Japan*. Routledge.

Part III

Global Dialogues About Crime and Destruction in the South

11

Revisiting Rosa: Eco-Bio-Genocide, Drug Wars, and Southern Green Criminology

Nigel South

Introduction

In this chapter I take advantage of reaching a stage in one's career where it is possible to look back and see influences, interests, and intertwining. Of course, we all tell ourselves stories that help us make sense of life retrospectively (Plummer, 2001) and there are inevitable lapses of memory and temptations of ego that probably shape most such accounts. However, using this device as a starting point here is not intended to provide any kind of autobiographical reflection that would stand up as 'history' or 'data' but simply to offer an opportunity to look back at some of my own research interests in two fields that I (and others) should always have considered as more deeply entwined, and that now—as the world has changed—are quite evidently profoundly interlinked. These interests are (i) the production, distribution, and prohibition of illegal

N. South (✉)
University of Essex, Essex, UK
e-mail: n.south@essex.ac.uk

D. R. Goyes (ed.), *Green Crime in the Global South*, Palgrave Studies in Green Criminology, https://doi.org/10.1007/978-3-031-27754-2_11

plant drugs (e.g. cocaine, heroin, marijuana), and (ii) harms and crimes affecting the environment (and by obvious extension, human and non-human populations that suffer as a result). For the purpose of connecting these thoughts to a Southern green criminology I want to start by revisiting a classic work by Rosa del Olmo (1937–2000). My focus here is twofold—on del Olmo's (see Fig. 11.1) work on drugs issues, but also on her sensitivity to the importance of environmental harms and victimisation.

This latter point creates the platform on which I try to build: first, to provide some commentary that updates del Olmo's work in relation to 'drug wars', and second in relation to environmental harms. Were del Olmo with us today, she would surely be engaging with a Southern green

Fig. 11.1 Rosa del Olmo in Mexico 1992 (*Source* South [2022]. Credits for the picture: Lina Torres Rivera, socióloga y criminóloga, catedrática de la Universidad del Sagrado Corazón, Puerto Rico)

criminology, the challenge of climate change, and how the continuation of drug cultivation and eradication, deforestation and resource extraction, are contributing to this and making the countries of Latin America pay a high price for meeting the demands of the consumer economies of the global North (Ayres, 2020; McSweeney, 2015a).

Rosa Del Olmo and the 'War on Drugs'

Professor Rosa del Olmo was 'a leading figure in the world of left-wing criminology. Pioneer of critical criminology in Latin America, feminist and inexhaustible social fighter and one of the most well-versed criminologists in the socio-political and economic analysis of drug trafficking' (Padilla, 2016). I first encountered some of Rosa's work in the early 1980s in the journal *Crime and Social Justice* (as it was then called, now *Social Justice*), notably her 1975 paper 'Limitations for the Prevention of Violence: The Latin American Reality and Its Criminological Theory' (del Olmo, 1975). This was, as Shank and Dod (1987, p. iii) indicate, a significant paper in the development of 'Latin American Critical Criminology'.

Around a decade later, del Olmo published her pathbreaking (1987) article, 'Aerobiology and the War on Drugs', which appeared in a special issue of *Crime and Social Justice*, dedicated to coverage of 'Latin American Perspectives on Crime and Social Justice'.[1] I am not, however, very sure about how widely known or acknowledged this work was in the UK at that time and I think it is possible (though my memory is not entirely reliable) that I only came across the paper on 'Aerobiology' in the early 1990s. The field of English-language sociological research on drugs in the 1980s was, as I recall, still thoroughly dominated by US sources, with the addition of key studies from the UK and Australia from the mid-late 1980s onward (Coomber et al., 2016). All of this meant, of course, looking at drugs issues—and many other

[1] Some of del Olmo's other wide-ranging contributions are noted in Goyes and South (2017, p. 176); she also published two book length critiques of drug policy and controls (del Olmo, 1990, 1992).

criminological topics—through a transatlantic prism (Aas, 2012). Even when adopting a political economy perspective that raised awareness of the harms, violence, and contradictions of a war on drugs that was targeted against producer countries in the global South, the discourse was distorted and the focus tended to remain aimed at the policies and impacts affecting the global North. This was brought home to me on an occasion in the early 1990s when I was speaking to an anthropologist who specialised in the study of various countries in Latin America. A distinguished scholar, with expertise in Marxist analysis of the economics and politics of these states, it seemed to me that we might discuss, for example, the significance of the cocaine economy or links between organised crime groups and political elites. However, I was trying to cross a disciplinary boundary or create a pipeline between two separate silos for there seemed to be little interest or keenness regarding such a discussion. To be fair, this did change several years later but I was struck then—and the project of trying to adopt a 'Southern' perspective emphasises this—that even when academic work arising in the global North was seriously concentrating on aspects of the global South, the way of 'seeing'—prioritising, interpreting, representing—could retain and even reinforce a Northern point of view.

Of course, I was also a product of my British and to some extent transatlantic academic culture, and any work I was associated with in these years was also limited in its engagement with what we now see as a 'Southern' perspective in criminology (Brisman et al., 2018; Carrington et al., 2016, 2018). There were, no doubt, many exceptions to such myopia that I did not encounter or have forgotten, but on this note I can mention one early signpost to important connections that I was involved with but perhaps did not learn enough from at the time. In the mid-1980s I was very fortunate to be working on ideas about different 'drugs economies' with Auld et al. (1984, 1986) who both brought a breadth of innovative thinking, derived from deviancy theory and Marxist political economy, to some joint writing projects. In particular, Auld et al. (1984, p. 4) is relevant here and noted how the model of 'development' promoted by ex-colonial powers in Latin America and Asia was conducive to the expansion of trades in plant drugs. This was not, of course, the stated aim, but programmes of 'modernisation' based on mineral extraction and agricultural cash-cropping, large-scale infrastructure schemes with minimal wage and welfare costs, accompanying social and cultural disruption, and the creation of national indebtedness to the global North, all encouraged alternative enterprises based on different varieties of cash crops and export strategies (see also Ayres, 2020, pp. 252–253). Nearly 40 years later it is acknowledged that 'illicit economies' have been drivers of socio-environmental change in Latin America, disastrously affecting Indigenous and Afro-descendent peoples, their lands and their rights (Dest et al., 2021).

Del Olmo begins her essay by noting the ubiquity of 'drugs' as a phenomenon and as a *word* in the 1980s: 'Men and women, rich and poor, old and young, Blacks, Indians, and whites, all know the word' (del Olmo, 1987, p. 28)—and it is indeed part of our cultural wallpaper (South, 1999, pp. 3–8). Alongside this cultural understanding, however, the distribution and sale of drugs are also, she says, '*viewed as crimes which harm international economic and sociocultural development of states and peoples* along with ecocide, contraband, illegal emigration, the counterfeiting of money, etc.' (del Olmo, 1987, pp. 28–29, italics in original). Del Olmo's analysis goes on to present a powerful critique of Reagan-era US drug policy which remains insightful and important, but for the present purposes it is significant that among the range of criminal acts, humanitarian issues, and criminological challenges that she refers to is the threat of *ecocide*. This was a clear outcome of the 'war on drugs strategy' adopted by the USA, insofar as it was both a war on domestic drug consumption, albeit with a biased concentration on some populations rather than others (Fordham, 2020; Patten, 2016), as well as a war on drug production outside the USA. The latter depended heavily on the adoption of 'programs of interdiction and eradication' that involved border seizures and what del Olmo (1987, p. 30) calls *aerobiological warfare*—the use of toxic chemical pesticides and herbicides 'disseminated with the air as a vehicle' (p. 32). Hence, as she aimed to demonstrate—referring to a definition of international crimes in the Declaration of Quito agreed by various Latin American presidents in Quito, Ecuador in August 1984 (del Olmo, 1987, p. 28)—'one can include among "the crimes which harm the economic and socio-cultural development of states and peoples" programs which have the characteristics of *crimes against the environment*' (my emphasis).

According to del Olmo (1987, p. 30), pursuing drug crop eradication programmes should be seen as the equivalent of a 'crime which has the characteristics of ecocide by virtue of making war with certain methods, systems, or prohibited weapons'. It is unclear from her list of references whether she was aware of the early condemnation of the use of chemical weapons in Vietnam as a crime of ecocide, as made by Professor Arthur Galston in 1970 at the 'Conference on War and National Responsibility' in Washington (see Gauger et al., 2012), and by Richard Falk (1973), though I would be surprised if she was not (she does cite news stories about Agent Orange and 'pesticide peril'). What is explicit is the comparison of the new drugs war being pursued by the USA to the earlier example of Vietnam as a military exercise in which weapons of choice include 'toxic chemicals, especially herbicides prohibited in their place of origin for causing poisoning, contamination of food, and serious environmental problems' (del Olmo, 1987, p. 30). In response to, and anticipating discussion that followed some years later on the prospects for the introduction of an international crime of ecocide (Gray, 1996; Higgins et al., 2013), del Olmo (1987, p. 30) notes the existence of a 1977 United Nations convention on environmental protection. This she says was to prohibit the 'use of agents that modify the environment for military purposes or other hostile objectives', so presumably this is the 'Convention on the prohibition of military or any hostile use of environmental modification techniques', dated 10 December 1976, signed 1977 (Schindler & Toman, 1988, pp. 164–169). In addition, she refers to 'agreements against ecocide, against the destruction of the environment, and more concretely, biocide'—which, quoting Karpets (1983, p. 44), is defined in her essay as 'the illegal, irreversible destruction of the environment, which implies the alteration of international security and affects the well-being and the health of not only the present generation of human beings but also future generations'.

While it does not seem clear what specific 'agreements' del Olmo had in mind that would apply with legal precision to either 'ecocide' or 'biocide', her comments were perhaps reflecting what she may have detected as some momentum toward recognising the importance of 'criminality against the environment'. Lopez Rey (1983) had, she noted, referred to the necessity of creating the phrase 'criminal protection of the natural world' in relation to the quality of life and the importance that it is gaining in international study' (del Olmo, 1987, p. 30). Del Olmo also invokes the idea of genocide which she applies to this context because 'crop eradication programs employ weapons with serious effects on the "quality of life," when life indeed remains'.

Regardless of some missing legal detail, all of this means that in the space of just a few lines del Olmo drew attention to key concepts that would come to underpin a 'green' criminology, and additionally to the work of two authors making pioneering early statements about the need for such thinking, yet which seem to have been largely overlooked in the history of green criminology—a further case of the amnesia of the field about precursors and precedents (Goyes & South, 2017).

This classic article also provides a number of examples of reports about the way in which the 'war on narcotics' was becoming a 'war on nature' and how it affected not just illegal crops but legitimate food crops (del Olmo, 1987, p. 28), as well as human and non-human victims. For example, detailed evidence gathered by the Colombian journalist, German Castro Caicedo, based on talking to drug crop cultivators revealed that 'the spraying is being done indiscriminately, not only against marijuana, but also against every type of food crop of the peasants, and their farms are being destroyed by the helicopters' (*El Mundo*, 23 July 1986: 1-A). In another case, an Indigenous leader in the Sabana Culebra region of Santa Marta, in northern Colombia, who described the effects of the herbicides was quoted in *El Tiempo* (29 June 1986): 'My brothers are dying there and what the doctors are doing, the medicines they leave, is not sufficient because this is an unknown and deadly epidemic. We believe that it is a poisoning produced by spraying with glyphosate because the symptoms are vomiting of blood, intense headaches, and shivering all over the body until death'.

Such accounts and evidence explain why 'genocide' is the final element of the concept del Olmo derives from her analysis as she proposes a 'transnational crime of broad scope which we can call *eco-bio-genocide*'. This, she argues, could be applied beyond the war against drugs because it involves the utilization of a whole complex of toxic chemicals, called pesticides, herbicides, or plaguicides, which are prohibited and/or restricted in the developed countries but have an unlimited market in Third World countries despite the well-known legal consequences (del Olmo, 1987, p. 31).[2] This is a rich and pioneering concept that has been under-used and for the purposes of a contribution to the present volume, one obvious if uncomfortable question to ask is whether the concept would have been more widely adopted if proposed from outside the borders of Latin American criminology.

Moving 'Beyond the West'

Del Olmo's classic paper clearly anticipated much that came to be central to a 'green' criminology, and when Piers Beirne and I were seeking contributors to a journal special issue that could outline the potential breadth of this emergent 'green' perspective, it seemed a perfect opportunity to ask Rosa to revisit her paper and write a sequel (del Olmo, 1998).

Despite Rosa's generous contribution—and the international examples and references made by other authors—that special issue did not really move much beyond an obvious Northern bias in the discussion and analysis offered. Our Editors' Introduction referred to concern with the 'destructive effects of human activities on local and global ecosystems' (South & Beirne, 1998, p. 147); my own paper (South, 1998, p. 226) suggested that part of the future agenda for green criminology should engage with '"new" topics of international and global importance'; and Benton's (1998) scholarly outline of challenges related to matters of 'rights and justice on a shared planet' mentioned growing awareness of 'ecological disasters flowing from the wreckage of oil tankers, the

[2] The terms 'third world' and 'developed/developing' are now subject to debate about their value and the fact that they reflect a global political analysis of some decades ago; they are also criticised for the way they can be taken to imply a hierarchy, failure, and a 'fixed' view of the world. Del Olmo used the terms when they were widely accepted and they are, of course, still in common use although terms such as the 'global South' have increasingly been used instead—although also not without critique.

chemical explosion at Bhopal, the desertification of parts of Africa, the destruction of the tropical moist forests' and so on. However, the essays in general reflected a view of the planet and its peoples from the perspective of the global North. This has, of course, been a recurrent problem with criminology's dominant viewpoints (Carrabine et al., 2020; Goyes & South, 2021) and no less so in the research literature on illegal drugs despite most of the production of plant drugs being based in the global South.

In a collection of essays on 'drug use beyond the west', Coomber and I (Coomber & South, 2004) had sought contributions that reflected contexts and cultural settings often neglected in 'Northern' literatures on drugs but that were greatly impacted by the legacies of colonialism and the contemporary power of international—by which is generally meant Western, and more specifically USA-led—prohibition and control agreements. The coverage included studies of cultures, rituals, and controls related to drug and alcohol use in Brazil, Bolivia, Argentina, India, Yemen, Mexico, Canada, Jamaica, and Kenya. Among the aims of the text was to set out ways that drug policy might learn from non-Western, non-problematic forms of drug use (Coomber & South, 2004, p. 23), recognising that what this would require was a challenge to the 'continuing power of the discursive understandings and operationalised systems of post-colonialisms [that] emphasise the need for the control of problems that are seen as residues of the past and impediments to "modernisation"'. The value of thinking about a 'post-colonial' approach at the time (however open to question the idea may be now, see e.g. Watson & Wilder, 2018) was that, as McLaughlin (2001, p. 214) suggested, as a 'mode of critique and challenge, it provincializes much of what passes for general theory and points to its Eurocentrism'. Subsequently, as Piers Beirne and I tried to develop the scope of a green criminology through editing a further collection of essays (Beirne & South, 2007), I began to think about my own Eurocentrism and the need to refocus on context and culture 'beyond the West'. In this I was influenced by the pioneering work of Reece Walters on food justice issues, and on the activities of corporations seeking to establish aggressive monopolies and push new bio-technologies such as genetic modification (Walters, 2004, 2006).

With the aim of exploring issues raised by Walters, I attempted to link thoughts on the definition and scope of a *green criminology* to an application of these ideas 'beyond' the West and the North and instead to the states and peoples that could be seen as victims of the biased and stacked system of resource exploitation. In this discussion I included a section on 'Globalisation: eco-bio-genocide as a transnational crime' (South, 2007, p. 233), summarising del Olmo's (1987) article and noting that, although not explicitly stated, her concern for those having to live with the residues and consequences of toxic crop eradication chemicals should be seen as a contribution to a 'green victimology' (Williams, 1996). It also seemed that an under-developed area at the time—in criminology at least—was what might be called, broadly, the 'theft of nature' (a topic investigated further in Goyes et al., 2016). More specifically, as I pursued this thought, a case study of a modern form of piracy—bio-piracy—emerged. My argument was that international law, corporate mobilities, and the exercise of intellectual property rights, all firmly rooted in Northern colonialism and capitalism, had enabled the persistence of a twentieth/twenty-first-century version of the piracy and pillaging of earlier privateers licensed by states and merchant companies (Shiva, 1998; South, 2007, pp. 240–241). I referred to this process as the 'corporate colonisation of nature' and explored the developments that had led to the 'patenting of biology' (all now far more thoroughly investigated and described in Goyes, 2018a, 2019). Astonishingly, applicable and effective regulation was and remains under-developed. That this was so in relation to governing genetic modification technology was partly related to the agility of multinational corporations switching jurisdictions for different activities, but also as a result of the phenomenon of 'regulatory capture', by means of which former law-makers and regulators join the management boards of companies in the sectors where they were previously engaged in oversight roles and suddenly demonstrate entirely sympathetic attitudes (Dal Bó, 2006). It also comes from the acquiescence of US administrations, particularly when the petitioner is a major and powerful corporation such as Monsanto. In 1986, Monsanto lobbied Washington regulators for rules that would govern the use of new technologies applied to the genetic modification of food. Monsanto was successful in seeing that the outcome was a set of regulations that fitted its aims and plans and, as Eichenwald et al. (2001, p. 1) describe, this was an outcome 'that would be repeated ... through three administrations ... If the company's strategy demanded regulations, rules favored by the company were adopted. And when the company abruptly decided that it needed to throw off the regulations ... the White House quickly ushered through an unusually generous policy of self-policing'. (See also Goyes, 2018b).

Rather importantly, among the 'pesticides, herbicides, or plaguicides' that del Olmo refers to as part of the armoury of drug crop eradication chemicals is one of Monsanto's leading products—glyphosate—marketed internationally since 1974 as 'Roundup'.

Drugs, Environment, and Climate Change

As Harvey (2021) shows, British colonisation of slave plantations in the Caribbean and Indian Ocean and the settler colonialism of North America were crucial points in the socio-genesis of environmental and climate changesand, as Castellino (2020, p. 577) observes:

> The contemporary climate emergency is directly traceable to colonial activities commenced on indigenous territories, continued under post-colonial regimes, with the active support (material and logistic) of the former colonial powers. These practices stimulated demand for 'products', treated territories as resource hotbeds, ... ignored the human rights of indigenous peoples.

Obviously, within this complex web of supply and demand, plant drugs (perhaps once legal, subsequently demonised and prohibited) have played a significant role—from the time of the opium empire of British colonial trade onward (Berridge & Edwards, 1987) as well as elsewhere (Ayres, 2020, p. 241). As Fordham (2020) points out, colonial interests in many parts of the world viewed these plants as important commodities to enrich their coffers. In particular, British, French and Dutch colonial powers conducted lucrative trade by producing opium, coca, and cannabis for export in their colonies in India, Burma, Indonesia,

Morocco and Algeria. In the case of Latin America, del Olmo (1987, p. 29) notes that the continent became 'a major producer of marijuana and the only producer of cocaine—with the great demand for both substances in the United States constituting the fundamental reason for their production and commercialization'. In her later 1998 article, del Olmo made the distinction between the 'two sides of the problem to be considered: first, the ecological impact of drug production itself; and second, the ecological impact of drug destruction' (p. 270). Several directions and questions follow from this, and here I explore how these have been investigated subsequently.

Environmental Costs of Drug Production

The focus of much research on 'drugs' from the 1960s onward was on consumption and forms of social control—the liberal attitudes and labelling theories of the period being relatively uncritical of or indeed curious about the sources of marijuana, cocaine, and heroin. Although critical studies and political economy analyses emerged in the 1970s (Coomber et al., 2015) there was little—if any(?)—reference to the environmental impact of the processes of production. Such linkages became clearer in the 1980s (e.g. Henman, 1985), but as Sesnie et al. (2017) noted, 'the role of drug trafficking as a driver of forest loss has only recently begun to be recognized (Nellemann, 2012)'. Yet, what Devine et al. (2021) call 'narco-trafficking' has been a significant cause of the consequence they call 'narco-degradation'—forms of land-use change involving wildlife trade and illegal logging that have undermined 'conservation governance' (Wrathall et al., 2020) and forcibly taken over smallholdings and cooperatives. Furthermore, as McSweeney (2015b, p. 8) describes: 'When drug crop cultivation moves into fragile ecosystems, ecological damage extends to environmental contamination', with millions of barrels of chemicals such as acids, ammonia, acetone, and kerosene, running off into the ground and local water sources annually, as farmers, processors, and law enforcement agents all go about their business in the continuing drug war.

Environmental Costs of Drug Eradication Policies

McSweeney (2015b, pp. 2–3) summarises the impact of the other side of the drugs production dyad, the consequences of control, as resulting in perverse harms to the natural environment, whereby 'drug crop eradication drives deforestation by progressively displacing drug farmers into new, more remote environments (Rincón-Ruiz & Kallis, 2013)'. This results in a cycle of enforcement attempts at interdiction which pushes cultivators and traffickers into the destruction of further bio-diverse habitats as new smuggling routes are created. The tensions and contradictions here mean that 'Support for drug supply eradication by the United Nations Office on Drugs and Crime (UNODC) and the UN Commission on Narcotic Drugs works in direct opposition to concurrent UN efforts to protect biodiversity, secure ecosystem services, ensure the rights of indigenous peoples, mitigate climate change, and promote sustainable development'. (McSweeney, 2015b, p. 3).

This is not unfamiliar, and elsewhere (South, 2016) I have commented on the way in which the imperatives of international organisations such as the World Trade Organization and the United Nations Environment Programme are logically and evidently in conflict, yet present to the world a message of 'joined up' cooperation. This is made possible by familiar means of reconciling—as Rayner (2012, p. 112) put it—'diverse principles' and 'constructive ambiguities' in ways that enable 'uncomfortable knowledge' to be accommodated and grounds for compromise and agreement to be found rather than pursuing questions about divergent motives, agendas, and worldviews. This bridging of vocabularies and missions reflects three key points:

> a desire to preserve 'business as usual'; the imperative that any need for fundamental change is downplayed; and, the final element, that all of this represents an assurance that *when necessary*, the fragmentation of positions on policy and practice, speaking and acting in contradictory ways, can all be made coherent through the articulation of higher level declarations of 'common interest', proving that 'something is being done'. (South, 2016, p. 49)

This argument acknowledges Ruggiero's (2015, pp. 99–100) suggestion that 'proper philosophical and political justification requires that partial concerns and factional gains be depicted as beneficial to the collectivity; therefore it entails agreed upon definitions of the common good and the identification of higher common principles'. In this case it seems that the highest attainment aspired to is drug crop eradication and—somehow—the environmental impacts are supposedly mitigated simply because they are known about. This is history repeating itself—as Shank and Dod (1987, p. xvi) wrote of the impact of the drugs war on Latin America in the 1980s: 'For the Reagan administration, the public condemnation of drug trafficking is demonstrably more important than the suppression of human rights violations and elimination of institutional violence.'

On the matter of violations, McSweeney (2015b, p. 6–8) notes that the environmental damage caused by the use of toxic defoliants and by aerial fumigation campaigns has been documented by 'several UN agencies and other multilateral bodies, such as the Organization of American States' (OAS, 2013), showing how, for example, the fumigation programme in Colombia lasted for decades and inflicted substantial 'collateral damages to wildlife, livestock, food crops, and the health of rural residents (Rincón-Ruiz & Kallis, 2013)' (see also Neuman, 2015).

Climate Change and Colonial Wars in Latin America

The transformation of Latin America's land-uses and economies as suppliers of commodities—whether legal beef or illegal drugs—to meet the high demand of markets and the corporate empires of the global North has made significant contributions to the warming of the planet. In a study focused on Central America, Sesnie et al. (2017) estimated that 'anomalous patterns of forest loss associated with narcotics-trafficking may account for between 15 and 30% of annual national forest loss in Nicaragua, Guatemala, and Honduras over the past decade'. In terms of policy aims to reduce climate change and maintain carbon capture capacity, McSweeney (2015b, p. 13) points out how UN and other programmes designed to encourage forest conservation (such as REDD+[3]) are undermined by the financial and coercive power of drug trafficker enterprises to bring about the conversion of forested areas to land for cultivation, or to build roads and airstrips for transport of precursor chemicals and final products (Dellinger, 2020). This transformation also diminishes the chances of ensuring 'long term, community-centered control and management of forests' (McSweeney, 2015b, p. 13).

In this context and in terms of the theme of Northern impositions on Southern environments, it is of historical note to refer to the climate *cooling* caused by the environmentally destructive wars waged by the Spanish colonialism of the sixteenth century. This was a holocaust against Indigenous peoples which according to Degroot (2016) 'forced many of the survivors into new settlement patterns and gruelling forced labour'. Colonisers 'disrupted indigenous ways of life by appropriating, and often transforming, regional environments', introducing rapidly thriving varieties of plants, animals, and pathogens. Significantly, 'Indigenous communities in hot, humid climates were especially vulnerable to outbreaks of Eurasian crowd diseases' and the figures for the resulting population decline in the Americas are startling—from 'approximately 61 million in 1492 to six million in 1650' (see also Dull et al., 2010).

A Southern green criminology needs a long view of the events of history as well as a wide view of the conditions of the present.

---

[3] Under the United Nations Framework Convention on Climate Change this is the programme of initiatives to reduce 'emissions from deforestation and forest degradation, as well as the sustainable management of forests and the conservation and enhancement of forest carbon stocks in developing countries' (https://unfccc.int/topics/land-use/workstreams/redd/what-is-redd).

Conclusion: Eco-Bio-Genocide and the Agenda for a Southern Green Criminology

As Carruthers (2008, pp. 4–5) notes: 'Concepts, discourses, movements, and policies emerge from specific political, cultural-historical experiences' and Northern 'conceptions of popular environmental activism … overlook many forms of action and consciousness that are central to the Latin American environmental experience, such as struggles for land and defense of traditional seeds or agricultural practice'. For criminology, combining a 'green' and a 'Southern' way of describing the Latin American 'environmental experience' is overdue and offers a wide horizon to explore.

This chapter has been a reminder of the connections between two of my past research interests. My limited attentiveness to these connections is perhaps a reflection of a wider lack of serious analysis given that, as McSweeney (2015b, p. 3) observes, 'the environmental costs of drug policies have been seriously underestimated and important stakeholders within and beyond the UN have been side-lined in drug policy debates' (see also Ayres, 2020, p. 239). The chapter has, however, also been about acknowledging a debt to Rosa del Olmo who made these connections earlier, more effectively, and who can be seen as a pioneer of what we are here, in this volume, calling a 'Southern green criminology'. These links are expressed clearly and succinctly in this passage from the conclusion to the article in which she revisited her original paper (del Olmo, 1998, p. 276):

> Both illicit drug production and drug crop destruction are important global issues, but the latter poses more questions, and should raise more concerns, because of the indifference of governments toward the major environmental and human health problems which pesticide use and misuse produce. ... These consequences are now very well documented and raise important issues both for the balance of drug-control policy, and for a criminology concerned with the environment.

In both contexts—and, of course, in drawing links between them—the concept of eco-bio-genocide should be revisited, for it is a powerful reminder of how it is the combination of human hubris, speciesism, and anthropogenic violence that is taking the planet to the edge of its climate 'tipping points' (Carrington, 2021).

References

Aas, K. F. (2012). "The Earth is one but the world is not": Criminological theory and its geopolitical divisions. *Theoretical Criminology, 16*(1), 5–20.

Auld, J., Dorn, N., & South, N. (1984). Heroin now: 'Bringing it all back home'. *Youth and Policy, 9*(2), 1–7.

Auld, J., Dorn, N., & South, N. (1986). Irregular work, irregular pleasures: Heroin in the 1980s. In R. Matthews & J. Young (Eds.), *Confronting crime* (pp. 166–187). Sage.

Ayres, T. (2020). The war on drugs and its invisible collateral damage: Environmental harm and climate change. In A. Brisman, & N. South (Eds.), *The Routledge International Handbook of Green Criminology* (2nd edn., pp. 239–259). Routledge.

Beirne, P., & South, N. (Eds.). (2007). *Issues in green criminology*. Willan.

Benton, T. (1998). Rights and justice on a shared planet: More rights or new relations? *Theoretical Criminology, 2*(2), 149–175.

Berridge, V., & Edwards, G. (1987). *Opium and the people* (2nd ed.). Yale University Press.

Brisman, A., South, N., & Walters, R. (2018). Southernizing green criminology: Human dislocation, environmental injustice and climate apartheid. *Justice, Power and Resistance, 2*(1), 1–20.

Carrabine, E., Lee, M., & South, N. (2020). 'Re-thinking public criminology: Politics, paradoxes and challenges' pp. In K. Henne & R. Shah (Eds.), *The Routledge International Handbook of Public Criminologies* (pp. 21–33). Routledge.

Carrington, D. (2021). Climate tipping points could topple like dominoes, warn scientists. *The Guardian*, 3rd June, https://www.theguardian.com/environment/2021/jun/03/climate-tipping-points-could-topple-like-dominoes-warn-scientists

Carrington, K., Hogg, R., Scott, J., & Sozzo, M. (Eds.). (2018). *The Palgrave Handbook of Criminology and the Global South*. Palgrave Macmillan/Springer Nature.

Carrington, K., Hogg, R., & Sozzo, M. (2016). Southern criminology. *The British Journal of Criminology, 56*(1), 1–20.

Carruthers, D. (2008). *Environmental justice in Latin America: Problems, promise, and practice*. MIT Press.

Castellino, J. (2020). Colonial crime, environmental destruction and indigenous peoples: A roadmap to accountability and protection. In M. Bergsmo, W. Kaleck, & K. Y. Hlaing (Eds.), *Colonial wrongs and access to international law*. Torkel Opsahl Academic.

Coomber, R., Moyle, L., & South, N. (2016). 'Reflections on three decades of research on "social supply" in the UK.' In C. Bernard & B. Werse (Eds.), *Friendly business: International research on social supply, self-supply and small-scale drug dealing* (pp. 13–28). Springer.

Coomber, R., & South, N. (2004). *Drug use in cultural contexts "Beyond the West": Tradition*. Free Association Books, London.

Dal Bó, E. (2006). Regulatory capture: A review. *Oxford Review of Economic Policy., 22*(2), 203–225.

Degroot, D. (2016). Did the Spanish Empire Change Earth's Climate? *Historical Climatology*, 2/29/2016. https://www.historicalclimatology.com/features/-did-human-brutality-trigger-preindustrial-climate-change

del Olmo, R. (1975). Limitations for the prevention of violence: The Latin American reality and its criminological theory. *Crime and Social Justice, 3*, 21–29.

del Olmo, R. (1987). Aerobiology and the war on drugs. *Crime and Social Justice, 30*.

del Olmo, R. (1990). *The hidden face of drugs*. Temis, Bogota.

del Olmo, R. (1992). *Prohibit or domesticate? The politics of drugs in Latin America*. Editorial Nueva Sociedad.

del Olmo, R. (1998). The ecological impact of illicit drug cultivation and crop eradication programs in Latin America. *Theoretical Criminology, 2*(2), 269–278.

Dellinger, A. (2020). How cocaine is contributing to climate change. MIC, 17th February, https://www.mic.com/impact/how-cocaine-contributes-to-climate-change-deforestation-in-central-america-19201947

Dest, A., McSweeney, K., & Sauls, L. (2021). Editorial: The role of illicit economies in transforming Latin America's Rural Spaces. *World Development*, special issue, July, https://www.sciencedirect.com/journal/world-development/special-issue/10B9CD78QC1

Devine, J., Wrathall, D., Aguilar, B., Benessaiah, K., Tellman, B., Ghaffari, Z., & Ponstingel, D. (2021). Narco-degradation: Cocaine trafficking's environmental impacts in Central America's protected areas. *World Development, 144*, 105474. https://doi.org/10.1016/j.worlddev.2021.105474

Dull, R., Nevle, R., Woods, W., Bird, D., Avnery, S., & Denevan, W. (2010). The Columbian encounter and the little ice age: Abrupt land use change, fire, and greenhouse forcing. *Annals of the Association of American Geographers, 100*(4), 755–771. https://doi.org/10.1080/00045608.2010.502432

Eichenwald, K., Kolata, G., & Petersen, M. (2001, January 25). Biotechnology food: From the lab to a debacle. *The New York Times on the Web.*

Falk, R. A. (1973). Environmental warfare and ecocide—facts, appraisal, and proposals. In T. Marek (Ed.), *Bulletin of peace proposals* (Vol. 1, pp. 80–96). Universitersforlaget.

Fordham, A. (2020) The war on drugs is built on racism. It's time to decolonise drug policies. *International Drug Policy Consortium.* https://idpc.net/blog/2020/06/the-war-on-drugs-is-built-on-racism-it-s-time-to-decolonise-drug-policies

Gauger, A., Rabatel-Fernel, M. P., Kulbicki, L., Short, D., & Higgins, P. (2012). Ecocide is the Missing 5th Crime against Peace, London: The Ecocide Project, Human Rights Consortium, School of Advanced Study, University of London. http://www.sas.ac.uk/hrc/projects/ecocide-project

Goyes, D. (2018a). *Biopiracy from a Green Criminological Perspective*, PhD Thesis, Oslo: University of Oslo. Faculty of Law Dissertations 119. ISSN 1890–2375

Goyes, D. R. (2018b). Corporate lobbying and criminalisation. *Crime, Law and Social Change, 69*, 401–419. https://doi.org/10.1007/s10611-017-9745-8

Goyes, D. (2019). *Southern green criminology: A science to end ecological discrimination.* Bingley.

Goyes, D., Mol, H., Brisman, A., & South, N. (Eds.). (2016). *Environmental crime in Latin America: The theft of nature and the poisoning of the land.* Palgrave.

Goyes, D., & South, N. (2017). Green criminology before 'Green Criminology': Amnesia and absences. *Critical Criminology, 25*(2), 165–181.

Goyes, D., & South, N. (2021). Indigenous worlds and criminological exclusion: A call to reorientate the criminological compass. *International Journal of Crime, Justice and Social Democracy, 10*(3), 115–128.
Gray, M. A. (1996). The international crime of ecocide. *California Western International Law Journal, 26*, 215–271.
Harvey, M. (2021). *Climate emergency: How societies create the crisis*. Bingley.
Henman, A. (1985) Cocaine futures. In A. Henman, R. Lewis, & T. Malyon (Eds.), *Big deal: The politics of the illicit drugs business*. Pluto Press.
Higgins, P., Short, D., & South, N. (2013). Protecting the planet: A proposal for a law of Ecocide. *Crime, Law and Social Change, 59*(3), 251–266.
Karpets, I. (1983). *Delitos de Caracter Internacional*. Editorial Progreso.
Lopez Rey, M. (1983). *International criminology*. Institute of Criminology, Universidad Complutense de Madrid: 152.
McLaughlin, E. (2001). Post-colonial criminology. In E. McLaughlin, & J. Muncie (Eds.). *The Sage dictionary of criminology*. Sage.
McSweeney, K. (2015a). The war on drugs is destroying the environment. *US News*, December 9th. https://www.usnews.com/opinion/articles/2015a/12/09/global-drug-policy-is-destroying-the-environment
McSweeney, K. (2015b). *The impact of drug policy on the environment*. Open Society Foundation.
Nellemann, C. (2012). Green carbon, black trade: illegal logging, tax fraud and laundering in the worlds tropical forests *Interpol environmental crime programme rapid response assessment*. United Nations Environment Programme, GRID-Arendal. www.grida.no. ISBN: 978-82-7701-102-8.
Neuman, W. (2015). Defying U.S., Colombia Halts Aerial Spraying of Crops used to Make Cocaine. *New York Times*, May 14.
OAS. (2013). *The drug problem in the Americas*. General Secretariat, Organization of American States.
Padilla, M. O. V. (2016). Rosa Del Olmo Pérez-Enciso (1937–2000). Pioneer of Latin American Critical Criminology. *Criminology. Paradigms and Theories on Criminality* @crimiparateorimaroctvazpa Book; Criminology. Paradigms and Theories on Criminality.—posts | Facebook.
Patten, D. (2016). The mass incarceration of nations and the global war on drugs: Comparing the United States' domestic and foreign drug policies. *Social Justice, 43*(1 (143)), 85–105.
Plummer, K. (2001). *Documents of life 2*. Sage.

Rayner, S. (2012). Uncomfortable knowledge: The social construction of ignorance in science and environmental policy discourses. *Economy and Society, 41*(1), 107–125.

Rincón-Ruiz, A., & Kallis, G. (2013). Caught in the middle, Colombia's war on drugs and its effects on forests and people. *Geoforum, 46*, 60–78.

Ruggiero, V. (2015). *Power and crime*. Routledge.

Schindler, D., & Toman, J. (1988). *The laws of armed conflicts*. Martinus Nijhoff.

Sesnie, S., Tellman, B., Wrathall, D., McSweeney, K., Nielsen, E., Benessaiah, K., Wang, O., & Rey, L. (2017). A spatio-temporal analysis of forest loss related to cocaine trafficking in Central America. *Environmental Research Letters, 12*, 5.

Shank, G., & Dod, S. (1987). Overview of 'Latin American Perspectives on Crime and Social Justice.' *Crime and Social Justice, 30*, 1–21.

Shiva, V. (1998). *Biopiracy: The plunder of nature and knowledge*. Green Books/Gaia.

South, N. (1998). A green field for criminology? A proposal for a perspective. *Theoretical Criminology, 2*(2), 211–234.

South, N. (Ed.). (1999). *Drugs: Cultures, controls and everyday life* (p. 164). Sage.

South, N. (2016). Free trade agreements, private courts and environmental exploitation: Disconnected policies, denials and moral disengagement. *International Journal for Crime, Justice and Social Democracy, 5*(4), 45–59. https://doi.org/10.5204/ijcjsd.v5i4.307

South, N. (2022). Arguing with Stan and Revisiting Rosa: Some thoughts on new criminologies in a changing world. *Papers from the British Criminology Conference, 21*, 6–19.

South, N., & Beirne, P. (1998). Editors introduction. *Theoretical Criminology, 2*(2), 147–148.

Walters, R. (2004). Criminology and genetically modified food. *British Journal of Criminology, 44*(2), 151–161.

Walters, R. (2006). Crime, bio-agriculture and the exploitation of hunger. *British Journal of Criminology, 46*(1), 26–45. https://doi.org/10.1093/bjc/azi049

Watson, J., & Wilder, G. (Eds.). (2018). *The Postcolonial contemporary: Political imaginaries for the global present*. Fordham University Press.

Williams, C. (1996). An environmental victimology. *Social Justice, 23*(4 (66)), 16–40.

Wrathall, D., Devine, J., Aguilar, B., Benessaiah, K., Tellman, B., Sesnie, S., Nielsen, E., Magliocca, N., McSweeney, K., Pearson, Z., Ponstingel, J., Rivera-Sosa, A., & Dávila, A. (2020). The impacts of cocaine-trafficking on conservation governance in Central America. *Global Environmental Change.*, *63*, 102098. https://doi.org/10.1016/j.gloenvcha.2020.102098

12

Colonialism, Knowledge, and the White Man's Burden

Rob White

Introduction

This chapter includes within its title the phrase 'White Man's Burden'. This is a highly loaded term. Its original meaning lay in its justification of imperialism and colonial expedition. It was the title of a poem by English author Rudyard Kipling in which he extolled the importance and virtues of United States control over the Philippine Islands (Kipling, 1899). This was framed in terms of a racialised notion of 'white' superiority over the non-white and was intimately related to the notion of the 'manifest destiny' of white European expansion and take-over of land and people worldwide.

For the purposes of this chapter, the expression is turned on its head. No longer a manifesto for white supremacy and imperialist endeavour, the play on words herein pertains to the obligations of the 'white man' to

R. White (✉)
University of Tasmania, Hobart, TAS, Australia
e-mail: r.d.white@utas.edu.au

D. R. Goyes (ed.), *Green Crime in the Global South*, Palgrave Studies in Green Criminology, https://doi.org/10.1007/978-3-031-27754-2_12

repair and respond to the horrors and dispossessions of previous epochs. Where does the burden lie when it comes to decolonisation? What role can those ostensibly privileged by colonial history play in changing its future in ways that reflect social justice and ecological wellbeing? Whose obligation is it to transform relations, and knowledge, in a manner than respects diversity, difference, and the sharing of power?

The term also has a slightly obtuse but nonetheless important connotation. With a last name like 'White', the phrase, perhaps more subtly, asks the question what can and should Rob White, individually and specifically, do about past and ongoing injustices perpetrated in the name of civilisation and hierarchical relations of power? As the chapter progresses, hopefully the reader will see that nuance and complexity, as well as context and biography, all come into play in unpacking both social and personal obligation.

This chapter is thus about 'obligation'. In it I assume historical and contemporary wrongdoing, much of which stems from or is related to white supremacist logics and practices. Parenthetically, it is worth noting that slavery and imperial conquest have not been the sole preserve of either 'whites' or 'Europeans' (e.g., consider the history of African and American kingdoms, the enslavement of the Hebrew tribes in Egypt, the exertion of Japanese power in Asia). But European imperialism has been fundamental and determinant in the shaping of contemporary national borders and the global political economy.

The importance of incorporating colonialism in green criminology analyses lies in the fact that exploitation of natural resources and people have and continues to be central to the colonial project. At the heart of this is appropriation of land. Control over land, whether by force, guile, or purchase, allows for the systematic despoilation and manipulation of the environment. Landscapes, rivers, habitats, and the subterranean depths are sacrificed on the altar of mineral wealth, flex crop profit, genetic modification monies, and dollars for the waste and contamination. Global power imbalances span centuries, right up to the present, and the impact and consequences shape billions of lives today.

The central focus of this chapter is on 'What is to be done?', but from the point of view of 'the coloniser' not the 'colonised'. To put it differently, while the march of decolonisation has happened over the course of the past century (evident, e.g., in the success of national liberation movements in Africa), and is still occurring globally in different forms (e.g., from nation-building at the country-level through to global recognition of Indigenous rights and identity), questions remain as to who is and should be involved in these processes, and how. This is not only a matter of specific activities and political movements; it fundamentally goes to the heart of knowledge and knowledge production itself.

Indeed, the extent to which decolonisation is occurring is subject to debate, given the changing forms of colonial rule over time (witness the granting of 'citizenship' to Indigenous Australians in 1967; but then consider the significant gap between Indigenous and non-Indigenous social indicators today). As with social life generally, there are ebbs and flows in the process, wins and losses, and setbacks to and expansion of rights and opportunities. The struggles continue, both over the terms of the accommodations between coloniser and colonised and in the power relations surfacing in periods of contestation. A case in point is the decolonial movement in academia, frequently situated in European and American universities that, in turn, function largely as neoliberal institutions. But more of that later.

This chapter explores these issues by considering two interrelated topics: identity and knowledge; and the intellectual field within which specific knowledge production occurs. In this chapter I wish to make a generalisation—about knowledge, about obligation, about social transformation—but will strive to do so in ways that incorporate the acknowledgement of the ambiguities, contradictions, and general messiness of doing so in an increasingly interconnected and globalised environment. The problems of essentialism and reductionism are touched on, but hopefully in a manner that does not end up with a relativistic notion that makes all knowledge 'equal' and thus each 'truth' equivalent. The intent is to question as well as assert, to differentiate as well as universalise. For a green criminology that purports to advance social and ecological justice, such considerations are fundamental.

Identity and Knowledge

Who 'we' are is intimately and ultimately tied up with self-knowledge, social connection, and the transfer of knowledge. Our concepts of the world are formed in concrete situations and contexts, and inform how we think, feel, and behave generally.

Cosmology and Knowledge

There is an inextricable link therefore between cosmology (understandings of the universe and our place in it) and knowledge (specific interpretations of the universe and our place in it), and these, in turn, shape our activities and behaviours. Ways of being, knowing, doing, and valuing (White, 2016) are socially constructed, historically located, and technologically driven. Explanations for lightning, for example, compete with stories of the gods at the advent of science explaining electricity. In a similar vein, a communal society does not have a crime of 'theft' insofar as private property does not formally or legally exist. Deviancy is defined more by the spiritual and ritual than material possession; that is, until property relations are introduced and new relationships between humans, and humans and objects, are imposed.

Much has been said in recent years about the cosmology of Indigenous peoples, especially in relation to the environment. This is reflected in United Nations initiatives and in global Indigenous movements for recognition and empowerment (United Nations Economic and Social Council, 2010). Typically, the knowledge at the heart of such representations includes concepts such as reciprocity, stewardship, custodianship, reverence, links between the human spirit and the natural life force, connection to land/country/nature, and collective spiritual identity (see e.g. Goyes et al., 2021; Graham, 2008; Robyn, 2002). Cosmology, as used here, refers to a totalising world view, an encompassing and encapsulating way of thinking, feeling, and being.

While cosmology is the abstract expression of lived relations and experiences, not all Indigenous people share in the same cosmology (nor do non-Indigenous people across the spectrum of social formations). The contingencies of history, invasion, hegemony, resistance, technology, and connection means that cosmology is epochal and intergenerational, selective and partial, with discernible changes as well as continuities over time. It is also refracted in biographical chance, as dictated by wider social circumstance. For instance, the 'stolen generations' of Indigenous children in colonial societies such as Australia and Canada were simultaneously deprived of family, language, country, and cosmology. Their relationship with land, kin, and culture was shattered by and rebuilt upon the edifice of the White Man's Burden. Intergenerational grief reflects the suffering caused by such intrusions. And the question of 'Who am I?' recurs in painful remembrance and existential crises.

The diversity within the colonial experience is one of the essential observations of the Australian Royal Commission into Aboriginal Deaths in Custody that reported its findings in 1991:

> The significance of land to Aboriginal people varies enormously as a result of colonization, the removal of people from their land and the removal of children from their parents. A rebuilding of family relationships is only now beginning to take place. Land that has become home to dispossessed people such as an old mission site or reserve land to which people had been removed has gained a significance in addition to, or in place of, a feeling for traditional lands now lost. There is also the extent of urbanization of Aboriginal people through these processes. All these aspects influence attitudes to policy with regard, not only to land rights, but also to appropriate forms of compensation that Aboriginal people feel are appropriate for survival into the future.
>
> It is important to appreciate that while there has emerged a pan-Aboriginal identity, the essential cultural characteristics of Aboriginal groups are intensely local. The cultural identities of local Aboriginal groups are often established and defined by the assertion of their difference from others. Local cultural groups are characterized by their traditions, beliefs and practices which confer upon them common interests in particular tracts of land in particular regions. This is so in spite of the fact that previous government policies demanded the forced removal of Aboriginal people from their traditional lands in many parts of Australia. (Johnston, 1991, p. 140)

Knowledge in the here-and-now and in relation to specific individuals and communities is thus always 'in motion' and 'in time'. It refuses to be reduced to biological and familial status, in the same way that 'politics' cannot be 'read off' from a person's skin colour, ethnic origins, or religious persuasion. We, the non-Indigenous, can learn much from Indigenous cosmology, but this does not mean that every Indigenous person carries within them the same worldview or ethical make-up. In this sense, the timelessness of Indigenousness (Graham, 2008) is 'out there' as well as 'in here' since it depends upon the integration of self and nature through acts of human agency. Not everyone is capable or willing to accept the invitation.

Hybridity and Identity

In the same way that knowledge is not reducible to 'race' and that personal cosmology is contingent upon a social context that often predates us, so too identity is not fixed and monolithic. The multiple dimensions of identity include, for example, the following attributes (White, 2013):

Chosen

I am who I say I am. (Agency in choosing identity.)

Bestowed

I am who others say I am. (Identities that entrench the sense of being Other.)

Simultaneous

I am more than one identity at the same time. (Bi-cultural and transnational identities.)

Strategic

I am who I am depending upon the circumstances. (Hybrid identities.)

Virtual

I am who I am as shaped by global telecommunications. (Virtual communities; transnational communities.)

An elaboration of this typology would explain the many ways in which identity is variable and complex, relational and presentational. For instance, it is capable of transcending the mundane in weird and wonderful ways—a Liverpool Football Club supporter identifies with LFC without ever being to Anfield or indeed Liverpool, yet there are bonds everywhere with others who similarly identify as being an LFC fan. Importantly, identity is strategic and is defined according to context and circumstance. Ascriptive characteristics, such as sex/gender and age, are difficult if not in some instances impossible to modify. But other dimensions of identity are malleable and changeable over time. Most importantly, we can be more than one person, at the same time.

Consider, for example, that in undertaking social research an individual may be positioned in several different ways within a community:

> in undertaking research in Fiji and Tonga on how communities respond to natural disasters such as cyclones, Johnston (2015) lived and worked in the small communities on remote islands. During one trip she brought her two young children with her as well as her partner. Her status as 'mother' and 'parent' opened up interesting opportunities to talk with local women about all manner of issues. Simultaneously, her presence as a 'white' 'Australian' 'researcher' also granted her certain status with the male leaders of the community. In essence, she lived a series of contingent and contextualised identities (Johnston, 2015). She was 'woman', 'mother', 'researcher' and 'outsider' depending upon circumstance and with whom she was interacting. Perceptions of her shifted according to situational factors but, in each instance, she had to be conscious of her own responsibilities and accountabilities, depending upon the nature of the exchange. Both identity and knowledge exchange was a continuous process of negotiation and dialogue. (White, 2022, p. 125)

Thus, identity depends upon social structure, situation, and agency. It changes for us and by us and can be used strategically in conscious efforts to gain knowledge as well as impart it. The reception of knowledge—its consumption, appropriation and circulation—is related to where it originates and who is saying what and for what purpose. Cosmology and identity may set the scene for knowledge production, but how knowledge is received and used frequently stems from pragmatism rather than idealism. This is evident in social constructions of 'really useful knowledge'. This refers to knowledge that is considered of value, for and by specific people, because of its practical relevance and associative familiarity.

Really Useful Knowledge

Over the years in Australia, mainstream criminologists have drawn upon official statistics to position Indigenous people as offenders, and usually as offenders with 'deficits' that need fixing with government support and/or coercion (for a critique of this, see e.g. Blagg, 2008). These approaches and viewpoints have subsequently been criticised by Indigenous intellectuals and critical criminologists who instead have emphasised a contextual understanding of colonialism, criminal justice, and Indigenous people (Blagg & Anthony, 2019; Cunneen & Tauri, 2016). Critics have also challenged the methodological basis for the originating analysis: namely, that there are no Indigenous voices in mainstream portrayals, nor is there control over the narrative by those most affected by it. Linked to this critique is the recent expansion of the Indigenous data sovereignty project, nationally and internationally, which, sociologically, represents a huge step forward in the statistical construction of reality (Walter & Suina, 2019; Walter et al., 2021). Socially, it represents a massive shift in the ownership and processes of knowledge production, distribution, and consumption.

Yet, in the midst of this assertion of, and in the reclaiming of, knowledge ownership, a number of ethical, practical, and theoretical questions remain. Specifically, what is the role of the 'white' scholar and researcher in the overall knowledge process? Who should say what, when and how, and on what empirical basis are and/or should judgements be made? In countries like Australia, Indigenous issues are intrinsic to doing mainstream criminology. For example, over 50% of young people nationally in youth detention are Indigenous (even though proportionately they only constitute 3–5% of the total Australian youth population). These issues, therefore, are an unavoidable fact of criminological study and teaching in this country. The same applies to green criminology insofar as resource exploitation globally is inevitably linked in some way to First Nations peoples.

Responding to these issues is highly contentious and subject to intense debate (see Anthony et al., 2021; Carrington, 2021; Moosavi, 2020; Watson, 2015). It consists of consideration of omission, bias, and exclusion as well as epistemological differences. Power is at the centre of the discussions, but how this is conceptualised and experienced is likewise debatable and debated. The oppression of the colonised by the coloniser is at the heart of some analyses; for others, the point of intervention is to accentuate the multiplicity of knowledges and the need for inclusion of diverse viewpoints. Many also emphasise that knowledge is not dichotomous (as in some portrayals of 'Western science' and 'Indigenous technology') and that knowledge systems can and do sit alongside each other (Goyes, 2019; Graham, 2008). It is not necessarily a zero-sum endgame in which one form of knowing always must trump another.

At the ground level, the response in Australia by Indigenous writers is: 'Nothing about us without us'. This is the key slogan advanced by Indigenous scholars and activists in recent years. But how does this work in practice? Some insight into the how was provided by my former University of Tasmania colleague Maggie Walter. For years Maggie and I have had informal chats about who should say what and who is responsible for speaking about contemporary issues of concern. As mentioned above, Indigenous issues are intrinsic to the field of criminology. In response to the ethics of commentary, therefore, when I wrote specifically about Indigenous young people or Indigenous communities, I would run the ideas and draft papers past Maggie. Her nod of approval was more meaningful that she could ever imagine. In these discussions, the crux of the matter came down to this: Maggie would ask (the following is my phrasing of remembered conversations, which Maggie has seen and acknowledged as accurate), *why should we (the Indigenous people) have to shoulder the burdens of colonialism and its oppressions when it is the 'white fellas' who are the problem? You (meaning 'white fellas') have a responsibility, too, in dealing with the issues that you have created. But this does not mean taking over the problem and making it yours—like some of your criminology colleagues. Rather, it means being sensitive to Indigenous perspectives and walking alongside Indigenous people as you address 'our' problems.* A lightbulb moment for me was when Maggie recently explained that 'Indigenisation is our project; Decolonisation is your project'. We work together, but on different fronts.

All of this is and remains very complicated. How we define specific terms, processes, and roles is an ongoing project with varying layers of uncomfortableness, confrontation, and listening. It certainly requires learning to hear differently and to maintain a critical stance in assessing any activity or intervention. For example, the appeal of Indigenous knowledge for many today, especially in the environmental social movements, lies in its cosmology and in particular the link between Nature and Human as expressed in the notion of Earth as Mother (see e.g. Huanca, 2019). What is abstract to others and reality to some is still somewhat theoretical to many. However, there are various ideas within the context of this cosmology that have application relevant to non-Indigenous institutions and people, including criminal justice.

The classic example of this in Australasia is 'family group conferencing' or as it has been institutionalised in Australia 'juvenile conferencing' (Cunneen et al., 2017). These forums stem out of considerations pertaining to restorative justice which itself has origins in various religious doctrines (such as elements of Judaism), secular theories (such as republican criminology), and Indigenous practices (such as Maori conflict resolution). The basic idea is that communities address transgression collectively so that they can repair the harm and reorient offender behaviour.

The appeal of these ideas, particularly in New Zealand/Aotearoa where family group conferencing originated, is that it presents as something 'practical' and 'applied' that can be grafted on to existing criminal justice systems and institutions. This is not an equal syncretic exercise. Rather, as emphasised by Maori writers such as Tauri (1999), the appropriation is a colonialist appropriation, one that is selective in degree, execution, and manifest outcome. Recent commentary on the granting of personhood to a river in Aotearoa/New Zealand under the presumed stewardship of local Maori makes a similar point (Monod de Froideville & Bowling, 2022); namely, that colonial law-making and criminal justice institutions do not fundamentally alter the colonialist relationship with those dispossessed and subject to colonial rule. Moreover, this is a present-day relationship, not a 'legacy' issue. The use of the past tense diminishes the continuing contemporary injustices.

By contrast, there are recent efforts in Australia to commission and learn from Indigenous people with expertise and technical know-how in regards to 'fire burning'. In this instance, the tables seem to have been turned. Rather than extracting Indigenous knowledge and institutionalising it for non-Indigenous benefit, the approach seems to be to allow Indigenous skills, knowledge, and understandings to come to the fore and to be 'owned' by them. The techniques are shared by Indigenous people with non-Indigenous people for the benefit of both. But the control is in the hands of the Indigenous people involved and the recognition of value is intrinsically acknowledged as Indigenous in origin. This particular 'really useful knowledge' has immediate application and acknowledged concrete results. This increasing acceptance of specific Indigenous knowledge and technology is, however, poles apart from the vociferous negativity of the previous Australian government in regards to the 'Uluru Statement from the Heart' written for and by Indigenous people which sets out a new agenda for Indigenous relations within the Australian polity. The latter is perceived by some as directly threatening white hegemony; the former is deemed relevant to fire prevention strategies and thus to protect the interests of all.

Symbolism and Engagement

Social change is occurring in Australia, but it is characterised by persistent negative continuities (such as the rampant over-representation of Indigenous people—men, women, and children—in criminal justice institutions, especially prisons) as well as important new transitional developments. On the plus side, virtually every event and most formal meetings in Australia (at least within universities and similar institutions), for example, begins with an 'acknowledgement of country'. At the University of Tasmania (UTAS) this typically consists of something like: In recognition of the deep history and culture of this place I wish to acknowledge the traditional owners and custodians of the land upon which we presently meet. I acknowledge the contemporary Indigenous community in Tasmania, and Elders past, present and future. Or: I want to begin by acknowledging the Indigenous people upon whose lands we stand—the *palawa* people of *lutruwita*—and their elders past, present and emerging. I thus acknowledge the contemporary Tasmanian Aboriginal community who have survived invasion and dispossession, and continue to maintain their identity, culture and Indigenous rights.

Personally, I have found it remarkable that in the last 20 years such acknowledgements have become routine and ever more detailed in regards to specific Indigenous groups linked to specific places. This *languaging of locality* is vital to the ongoing decolonial project and it is significant that often it is local municipal councils, dominated by non-Indigenous people, that have led the way in expressing Indigenous connection to land in this manner. The symbolism is not insignificant, as is the growth of public interest in 'our' collective 40,000 to 60,000-year history (which many Indigenous people wish to 'share' with other Australians) and in Indigenous languages (which many Indigenous people are only now learning for the first time due to the impact of the 'stolen generations' on their culture and language acquisition).

Decolonisation needs to occur everywhere that colonisation has taken place. This includes countries such as the United States. While 'acknowledgement of country' is now common in places such as Australia, Canada, and New Zealand/Aotearoa, it is less so in the heartland of contemporary global capitalism. At the November meetings of the Division of Critical Criminology and Social Justice (DCCSJ) in 2019, I opened proceedings this way:

> In recognition of the deep history and culture of this place I wish to acknowledge the traditional owners and custodians of the land upon which we presently meet—the Ohlone people.
>
> I acknowledge the contemporary Indigenous community, and Elders past, present and future. The Bay Area has become one of the largest populations of Intertribal Indians in the country with people coming from communities in the Southwest, Great Plains and Eastern Woodlands areas.

> I recognise that each year on Alcatraz Island, the International Indian Treaty Council organises and sponsors an Indigenous People's Thanksgiving Sunrise Gathering, a celebration of Native American culture and activism.

At the same meeting, I put forward the idea that the DCCSJ should open each formal meeting with a similar statement of recognition and acknowledgement. The political and cultural context of the USA is, however, very different to experiences outside of this country. For many American citizens, such statements still seem to strike them as foreign and somewhat strange and unsettling, probably in part due to the predominance of slavery as another issue of ongoing contemporary resonance and relevance.

Things are steadily changing in places such as Australia as wider collectivities such as universities, non-government organisations, and local councils adopt progressive policies as well as express their support for Indigenous people via symbolism and rhetoric. Much more needs to be done, but in the midst of much 'bad news' (again, the massive over-representation of Indigenous people in regards to not only criminal justice but child protection, poverty, illiteracy, poor health and life expectancy, etc.) there are nonetheless promising developments within my own 'backyard'. For example, UTAS was one of the first in the country to appoint an Indigenous criminologist.

More than that, UTAS offers broad first-year subjects on 'Indigenous Life Worlds', written, taught, and assessed by Indigenous scholars, supported by non-Indigenous colleagues. I had the fortune of co-teaching a section of one unit with Michael Guerzoni, our Indigenous criminologist. All the background material and readings were initially chosen from *palawa* (Tasmanian Indigenous) sources, with non-Indigenous articles supplied only as supplementary and sparingly to this. Michael and I took turns recording the content of short lectures, but the lead was provided by him, and the overall unit/course was distinctively Indigenous. Distinguished Professor White was 'Rob' and my contributions were only a small part of the sharing of knowledge that went on. For me, this was an exemplar in collaborative production, one that was intimately shaped and 'performed' by Indigenous people in lead roles. This required biting my tongue at times so that others could lead, and recognising that a variety of sources (e.g., artwork, newspaper clippings, oral histories, site visits for students) is beneficial in bringing this Indigenous knowledge to life.

Decolonisation is not only or simply 'out there' in the form of nice sounding words and exhortations within and by the academy to acknowledge the Indigenous subject. It is also 'in here' insofar as obligation extends to each of us, individually, to do our part when and if appropriate. This can be confronting and challenging. It means stepping back and truly listening. It means reading unfamiliar texts and grappling with unfamiliar ideas and ways of relating. Only through humility and giving up certain privileges can the decolonising project work at the hands of the non-Indigenous.

Yet, having said this, it is also important to not conflate collaborative knowledge production with uncritical thought. As discussed in greater depth elsewhere (see e.g. Ciocchini & Greener, 2021), one should not reify or render unproblematic knowledge that is defined or described as 'Indigenous' simply because it is so defined and described. Rather, the point is to assess such contributions on criteria such as whether it is emancipatory, representative, accurate, and reliable in the contexts within which it is constituted. Critical thinking does not stop at the decolonising door.

Nor is Indigenous knowledge and technique static and unchanging in its practical or applied dimensions. While cosmology speaks to how the universe makes sense of itself and the values that inform relationships, real world experiences always involve syncretic processes and contradictory and ambiguous outcomes. For instance, consider the idea of 'traditional' in the context of fishing and Indigenous people (see White, 2008). The word 'traditional' can refer to dimensions such as: *who* specifically (e.g., particular group or tribe); *how* specifically (methods, techniques, and technologies); and *where* specifically (traditional fisheries for particular coastal groups). Conflicts can arise when modern technologies are utilised for what used to be subsistence fishing. The use of motorboats, nets and fishing rods, and sonar equipment, for example, allow for overexploitation to occur. Moreover, overexploitation may be generated in and by the new methods of production themselves.

For example, on the one hand, the mobility, range, and efficiency of 'traditional' fishing are all enhanced through modern methods and technologies. On the other hand, these technologies generate the need for cash to supplement subsistence (e.g., buying the boat and petrol for the boat). The net effect is pressure to fish beyond immediate consumption needs. Hence, how knowledge and technique change over time warrants close consideration as do the ways in which they transcend time and space.

Intellectual Field and Knowledge Production

Knowledge production is not only never socially neutral, it always occurs in specific social contexts. Our obligations exist not only in regards to counter-hegemonic knowledge, but in addressing the conditions under which knowledge is produced and transferred. This chapter concludes, therefore, by making a few observations about contemporary academia and the nature of intellectual labour. In keeping with the green criminology theme, the section begins by reviewing environmental sustainability initiatives in the higher education sector.

Sustainability in the Higher Education Sector

In recent years, many universities have been actively dealing with climate change at an institutional level, frequently linking their actions to the United Nations Sustainable Development Goals (in particular, Goal 13 on climate change). One of the most celebrated universities in advancing policies and practices of sustainability is UTAS, which recently was named the '**Sustainability Institution of the Year**' in the 2021 Green Gown Awards Australasia. The Green Gown Awards support change towards best practice sustainability within the operations, curriculum, and research of the tertiary education sector. UTAS is additionally one of only two universities in Australia that is certified carbon neutral. This means it has closely audited and reduced emissions from all aspects of its operations and has offset its remaining emissions. UTAS has held this status since 2016.

Divestment from fossil fuels is another crucial aspect of meaningful climate action at universities. In 2020, UTAS announced its divestment from fossil fuels. The cornerstone of a transition away from investing in fossil fuels is a combination of negative screening (no new investments in fossil fuel related industries), positive screening (investment in renewable energy and ecologically sustainable industries), and the phased withdrawal of present investments in fossil fuel related industries and activities.

The university's support for climate action is evident in other ways as well. For example, its 'Sustainability Committee' is engaged in a wide range of activities. It is currently developing guidelines for a climate sustainable food culture at UTAS covering procurement, preparation, waste management, growing, vending, and food systems education. 'Education for sustainability' principles are also being integrated across all areas of the university. UTAS's sustainability policy will provide a strategic framework for actions into the future, guiding, for example, what the university does in regards to air travel carbon emissions, how it embodies the UN Sustainable Development Goals, and how it implements a circular economy action plan.

Viewed from the outside, UTAS looks very impressive indeed as this kind of 'greening' of higher education continues apace across the sector. However, viewing sustainability solely in institutional terms misses the human element of university life. Consideration of this provides a very different picture of universities today, one which also has implications for knowledge production, transmission, and reception.

Educational Workers in Social Context

Recent developments in the higher education sector in countries such as Australia, the United States, and the United Kingdom are impinging upon the work practices of educational workers in severe and negative ways and thus the nature and content of intellectual labour. Much of this revolves around the intensification of labour and changes in and over the labour process. In the UK, it currently also includes severe cuts to pensions and wages. The 'educational worker' is also socially constructed through its own peculiar 'professional' ideology and culture (the habitus of higher education). This, too, has implications, both for the future of educational work and the solidarity required to resist threats to the autonomy definitive of intellectual endeavour and thus the potential fulfilment of its progressive social obligations.

Working in universities changed remarkably for many of us sometime around 2010. From wonderful collaborative and friendly environments, we flipped over, very quickly, into metrics-centred, hyper-competitive, dog-eat-dog environments in which no one (or at least fewer of us) was happy. We still performed exactly the same work—research, teaching, administration, and community engagement—but the context had been radically altered. It has been a time that privileges top-down corporate management, as neo-liberalism has captured the university sector and government funding has been significantly scaled down. The word 'intellectual' has practically been diminished and in effect banished. We are not here to think and to ponder. We were here to produce, and to produce what the lesser academics (now our administrative bosses) define as the most productive 'research' outputs. The inherent individualism of academia has come to the fore and the intellectual project is itself under threat of abandonment.

The proletarianisation of academia is evident in the intensification of labour (increased volume of work), the increase and rigidification of the division of labour (specialisation into teaching, research, and administrative tasks at varying levels of responsibility), and the routinisation of high-level tasks (expert intervention and codification of existing tasks). In practical terms, this translates for some intellectual workers into the reduction of autonomy (immediate control over the labour process), the de-individualisation of skills and qualifications (fragmentation of tasks), and the downgrading of status (income levels in relation to volume of work performed).

The casualisation of the academic workforce is one manifestation of these changes in the overall labour process within academia. Short-term contracts, and less-than-full-year permanent positions, make intellectual workers financially vulnerable in their own lives in the same moment that they represent cost-savings for the employer. So, too, the massification of intellectual production in the form of online courses consisting of pre-packaged materials that 'anyone' can teach provide a ready income stream at relatively low cost for institutions of so-called higher learning. Institutional responses to COVID-19 have served to intensify these processes and consolidate these trends. They enable and ensure the transferability of content without paying the full costs of the living labour embedded in the educational commodity. Workers become ever more interchangeable and thus devalued for their specific contributions. Homogenised product and generalised production processes translate into less need for specific skills, knowledge, and human input. The result is a perversion of the intellectual mission and enhanced exploitation of the educational worker. As discussed shortly, it also impacts upon definitions of 'sustainability' and work conditions pertaining to this.

What 'counts' in institutional terms has consequences for knowledge production and the career prospects of knowledge producers. The intellectual landscape in criminology is contingent upon what research gets commissioned; who gets published and by which journal or publisher; whether specific forms of evidenced-based research dominate to the exclusion of other types of research; which departments or researchers are invited to tender for consultancies and research; and the extent to which critical criminologists are marginalised or embraced institutionally, professionally, and politically.

The petit bourgeois positioning of intellectuals within academia also forms part of the cultural field within which we work—that is, the emphasis on 'personal autonomy' (as professionals), 'individualisation' (in relation to achievements and careers), and 'privatised knowledge' (authorship is exclusionary—this knowledge is 'mine'). There are colleagues who write about neo-liberalism and various aspects of social life, and who publish quite successfully on these 'topics'. Their careers are furthered by writing about the very thing that structurally tears at the walls of academia. Yet, they themselves epitomise the neo-liberal subject. They are detached from students and colleagues, unwilling to join collective struggles or become members of unions, and if in positions of authority, carry out departmental negotiations behind closed doors with the institutional power-brokers. Their 'work' is entirely separated from any analysis of their 'conditions of work'. They are what they write about. They are in fact their own worst nightmare.

Our labour as intellectuals is mobilised and shaped by institutional contexts and structural forces. Our future—personally, collectively, societally—is always reflective of social structure, and how we interrelate with this. Hence, the vulnerabilities of the early career academic are real and meaningful, as is action by established academics in defence of their interests and prospects.

Sustainability?

In the midst of all this, academic and professional staff are nonetheless doing exciting and innovative work in their areas of expertise to combat global warming and promote climate justice. Simultaneously, institutions are rejigging their operations and rewriting their strategic plans to incorporate sustainability principles and practices. We all seem to be on the same page. Yet, there is a disturbing disconnect between the 'ordinary' work of the institution and the publicly pronounced commitments to sustainability goals. At the heart of this disjuncture is working conditions. Fundamentally, 'sustainability' at the institutional level must also incorporate the concerns and wellbeing of those who work within the institutions. In other words, sustainability goals only make sense when accompanied by careful consideration of the 'human factor' and not simply the institutional financial bottom line.

This is important in several respects. For example, the structure of decision-making and its basis have great impact on educational workers. When decisions are made top-down (rather than being collegial) and informed predominantly by economic criteria (rather than pedagogical mission), then this inevitably has effects on staff morale and performance. Poor working conditions that contribute to stressful and anxiety-producing environments suck the energy and commitment out of academic and professional staff. It is an alienating experience that has multiple negative consequences—for individuals, for programmes, and for universities.

Higher education in places such as Australia has seen extensive growth in the bureaucratic layers of universities and, simultaneously, the growing casualisation of the academic workforce (Kniest, 2021). Job losses have affected academic staff greatly, all the while that new hiring has privileged non-academic positions. The precariousness of academic tenure, for academics at all levels within the educational enterprise, means that the power of the centre has been considerably enhanced. It also means that educational workers have had to work harder to keep their positions. Control within and over the labour process in academia is primarily exercised 'from above' directly through executive fiat and in the form of predetermined metrics designed to measure outputs and engagements. Each of these further diminishes the traditional and ordinary decision-making of professionals and academics as they carry out their work tasks. Workloads have increased, and this has been exacerbated by the demands of responding to COVID-19 health and safety initiatives (including, e.g., the extensive use of zooming and other technological means of delivering lectures and services).

University treatment of staff has tended to reflect pandemic responses that put more pressure on staff to perform, in ways that save money via reliance upon 'home-based' work and technologies, and so also fulfil the objective of reducing the costs of 'sustainability' (at the institutional level—not for those working in their own homes). The individualisation of responsibility for making things work is also evident in exhortations to use public transport in lieu of reduced car parking spaces at the institution (except, of course, for higher level executives). Again, the reduction in allocated car spaces means that universities can signal their virtue in regard to sustainability objectives, but this comes at the cost of those working for the institution.

In a similar vein, while close attention to internal production and consumption processes provides a template to reduce waste and reliance upon fossil fuels, the overarching strategic plan of an institution may well generate additional emissions. For example, while UTAS has a well-developed sustainability programme that underpins its operational activities, the university has also embarked on a plan to move its main campus in Sandy Bay into the central business district of Hobart. This costly exercise has many elements. Notwithstanding the commitment to use the most efficient building materials and methods, much of this new development will require refurbishing heritage sites and taking over large parts of the city centre, that will then be linked by walkways. The proposal includes the construction of mini-parks and greater reliance upon public transportation although, in practice, there are significant logistical, policy, and political (especially in regards to local government and state government) obstacles and clarifications needed.

Meanwhile, the green spaces of the Sandy Bay campus will be transformed into other purposes including for commercial benefit. The move will require considerable new buildings as well as impacting greatly upon educational workers (for whom offices will be mainly open-plan) and students (who will struggle to find either a central hub in which to congregate or suitable grassy areas in which to sit). Sustainability measured in carbon emissions and infrastructure pollution may well reduce over time. But the danger is that the institution loses its heart and soul in the process. The 'virtual' university offers some opportunity—but it also precludes the building of human community and social life essential to the intellectual project.

Such a move into the city is simultaneously both confirming and undermining of the claims to 'sustainability'. Like the idea of 'green prisons', it is not only the physical sustainability that counts. For prisons we need to reduce numbers and orient toward rehabilitation (White & Graham, 2015). For universities, we need to pivot back to teaching as 'live' and 'personal' social interaction and to focus on intellectual production and consumption—otherwise the 'green university' is a misnomer since the cost–benefits of education are being driven by budget considerations (e.g., open-plan offices) rather than pedagogical and scholarly objectives (e.g., spaces to accumulate 'stuff', congregate *en masse*, think on our own, and interact with others).

The pursuit of knowledge is thus being submerged in the scramble to reinvent the mission in times of fiscal crisis. And yet, more money keeps going to poorly thought out infrastructure spending (e.g., old hotels and heritage buildings that are costly to refurbish), and bad and fanciful design (e.g., lack of adequate parking, public transport, and the creation of little mini-parks), resulting in a loss of cohesion (e.g., districts that are too segmented and fundamentally unattractive to visit or 'hang around' in). New development becomes a money-sink, in ways that further alienate students and staff from core business, and ultimately will cost in enrolments, morale, and the basic 'joy of learning'.

Decolonising Knowledge in Problematic Circumstances

These developments in higher education, and specifically at UTAS, are bound to have an impact one way or the other on both the Indigenisation of the academic curriculum and the decolonisation of knowledge generally (see also, Kidman,2020). It also makes the non-Indigenous obligation to social justice both more intense and more problematic. Let us talk, for instance, about "speaking truth to power", particularly for social scientists and those in the humanities. Speaking truth to power is or should not be simply reserved for selective public issues. It is relatively easy for me to join students and go on strike for climate, to support or be part of the march for women, to be an ally of Indigenous demands for self-determination. But that is not the same as challenging the powers that dominate our own workplaces. Suddenly it becomes a lot harder, because basically the vulnerabilities are real. And the thing about speaking truth to power is that if you are vulnerable, it is indeed much harder to act.

One of the most publicly performative aspects of universities in recent years in Australia has been their active support for Indigenisation of the curriculum and the hiring of Indigenous academics. In and of itself, this is a welcome development, and the fight by Indigenous intellectuals to forge these new agreements, arrangements, and innovations should be fully acknowledged. However, the context within which this is occurring is terribly difficult for most educational workers. It adds additional demands to already over-worked academic staff as they are tasked with revamping their units and courses without additional teaching relief. Also, there is a tendency for the newly hired Indigenous academics to be loaded up with extra duties on top of their basic discipline responsibilities. 'No knowledge about us without us' is translating concretely into onerous demands being placed on specific individuals, under the rubric of progressive social change.

As previously discussed, this is occurring in a work environment characterised by top-down imposition and executive fiat (frequently involving fake or meaningless 'consultation' with staff). In other words, the tailoring of academic work to fit specific institutional purposes is experienced as deskilling where, as Braverman (1974) pointed out years ago, there is use of a monopoly over knowledge to control each step of the labour process and its mode of execution: 'the essential element is the systematic pre-planning and pre-calculation of all elements of the labor process, which now no longer exists as a process in the imagination of the worker but only as a process in the imagination of a special management staff' (Braverman, 1974, p. 119).

Structurally, these developments are reinforced by making academics too busy to do what they profess they want to do—namely, engage in intellectual labour. This is what has previously been referred to as the 'busyness of business' (White, 2001). The conditions of academic work are intrinsically linked to issues of power/knowledge. The ironic question for a criminologist is whether these conditions of labour are leading to the production of overactive, but docile, bodies (see Foucault, 1977).

And thus, it is here that we see yet another contradiction, another ambiguity pertaining to colonialism, knowledge, and the White Man's Burden. For Indigenisation of the curriculum is an essential but not sufficient condition for social transformation. The other needed ingredient is decolonisation and thereby the active embrace of new knowledges and new ways of thinking, feeling, and being. Yet, institutional requirements and skewed work conditions present as barriers to the profound personal and group changes that ought to accompany this commitment to decolonisation. Moreover, they channel efforts into predetermined pathways that invite routinisation and (somewhat ironically) desensitisation. The 'acknowledgement of country' words are spoken, but their meaning becomes lost in the institutional ritual, particularly when the institution disrespects those working hardest for it.

One is reminded of the observations of American political activist Angela Davis:

> I have a hard time accepting diversity as a synonym for justice. Diversity ... is a strategy designed to ensure that the institution functions in the same way it functioned before, except now that you have some Black and Brown faces. It's a difference that doesn't make a difference. (Angela Davis, quoted in Eckert, 2015)

These processes can also drain the subject matter of its potential radical content and obscure the need for more profound forms of dissent (e.g., abolitionism and why this is substantively important to decolonisation). The undermining of the intellectual project is accomplished through such means. Talking the talk (of 'acknowledgement', of 'reconciliation') is all right from the institution's points of view, but walking the walk of publicly challenging criminal justice authority and systems is less so (particularly if university income is tied to the training of police, correction staff, and justice officials). In 2021 in Tasmania, academic staff at UTAS were 'reminded' at the start of the state election campaign about how to behave in public. The message was clear: 'be careful what you say and who you offend'.

In the light of current institutional pressures and limits, many academics are hunkering down to individually manage the dilemmas now permeating the sector. An apparent laziness or lack of engagement is a natural response to an institutional machine that will chop you up and spit you out. The more the bosses ramp up the pressure, the more people go quiet. Of course, without tenure, why would contract or casual staff protest or make themselves a target because of their political activism? On the contrary, they work even harder, saying yes to things they have no capacity to do in their paid time in the faint hope that that extra work will lead to a 'real' job and security. They are being forced to compete against and climb over one another to get to the top of the job application pile. This is a constant process that occurs all the time, and it is moving ever upwards the academic food chain.

In response, activist intellectuals argue that we need to radically engage within the university as part of the broader struggle for social justice. We need to join and be an active member of our union, share knowledge and know-how with our colleagues, and strive together for a vision of a university system that is democratic, engaged, truthful, creative, and sustainable (Connell, 2019). The doing of intellectual labour, including the work of decolonisation, does not occur in a social vacuum. Nor does it exist solely in the mind. It is demonstrated at the level of practice and in speaking truth to power even when that exposes one to vulnerability and the likelihood of retribution. The 'intellectual' is defined precisely by their autonomy in voice and action (Connell1983Said, 1994; Wright Mills, 1959). Defending this autonomy means acknowledging and responding to a changing workplace that reduces the dignity, capacity, and potential of educational workers.

The decolonisation project is not separate from these struggles. In each case the conflict is over purposes and power, sectional interests, and divergent viewpoints. A social institution based on exploitation and hierarchical domination mirrors relations of oppression experienced by others. The fight is not the same, but the struggles are similar in that each likewise involves conflict over knowledge production, consumption, and transference. The position and circumstances of educational workers must be an integral part of the discussions about decolonisation and Indigenisation since this has a major bearing on what kind of intellectual labour is likely or possible, by whom and for whom. It also shapes agendas and social constructions of what is 'safe' and what is 'dangerous' when it comes to the substantive content of the decolonisation process.

Conclusion

We cannot choose our birth or in the main our ascribed status (e.g., I am a white, older male)—but we *can* choose our *politics* (by writing about and acting on matters of inequality, discrimination, oppression, and subjugation), remembering, as well, that identity is complex and has multiple dimensions. Important struggles dominate the political landscape today (climate change being the greatest existential threat), and we need strength, determination, and collective effort to make a significant difference. We need to do this separately as appropriate (e.g., Indigenisation by Indigenous people; decolonisation by white people), and together as allies and progressive activists on matters of binding concern (e.g., global capitalism, climate change).

This particular 'white man' is defined by many personal experiences as well as by my social positioning from birth. These include being born in Germany, growing up in Canada, spending most of my life in Australia, and having spent the last 23 or so years living in Tasmania. They include living in a town of 6000 people as a teenager and commuting to a high school in Nova Scotia that had the term 'rural' embedded in its title (i.e., West Kings Rural High School). They include being the 'son of a preacher man' (I used to jokingly say 'My Father, my father' when speaking about my dad), an affiliation with formal religion that was sundered early on, although the values have traversed the years. They include being a migrant and struggling over many years to define who I was and whether and how to describe my national affiliations (which now also helps to explain my so-called 'mongrel accent'). Given these backgrounds, the imposter syndrome was inescapable for me, including bouts of self-doubt as I entered new intellectual adventures and social justice projects. Not all is as it seems, including for someone now 'Emeritus Distinguished Professor'.

Knowledge production is born out of biography, as well as intellectual endeavour and hard work. And it is always partial, ambiguous, contradictory, and contentious. There is Truth. But we come to this Truth by seeking perspective, anticipating and grappling with dissent, confirming facts, and maintaining dialogue. We also know that Truth hurts especially for the dispossessed and the traumatised, but also for those on the other side of the fence, for people like me who may be disturbed and threatened by it. Yet through the pain and the exposure comes greater potential for self-knowledge, a needed social rupture, and revolutionary transformation. The White Man's Burden is basically about debt and obligation and accepting our role in decolonisation. For out of this, new futures do indeed become possible.

References

Anthony, T., Webb, R., Sherwood, J., Blagg, H., & Deckert, A. (2021). In defence of decolonisation: A response to Southern Criminology. British Society of Criminology, Blog post.

Blagg, H. (2008). *Restorative visions: Crime, aboriginality and the decolonisation of justice*. Hawkins Press.

Blagg, H., & Anthony, T. (2019). *Decolonising criminology: Imagining justice in a Postcolonial World*. Palgrave Macmillan.

Braverman, H. (1974). *Labour and monopoly capital*. Monthly Review Press.

Carrington, K. (2021). *What needs de-colonising? Metropolitan thinking in criminology*. British Society of Criminology.

Ciocchini, P., & Greener, J. (2021). Mapping the pains of neo-colonialism: A critical elaboration of southern criminology. *The British Journal of Criminology, 61*, 1612–1629.

Connell, R. (1983). Intellectuals and intellectual work. In R. Connell (Ed.), *Which way is up? Essays on class, sex and culture*. George Allen & Unwin.

Connell, R. (2019). *The Good University: What universities actually do and why it's time for radical change*. Monash University Publishing.

Cunneen, C., & Tauri, J. (2016). *Indigenous criminology*. Policy Press.

Cunneen, C., White, R., & Richards, K. (2017). *Juvenile Justice: Youth and Crime in Australia*. Oxford University Press.

Eckert, M. (2015). *Civil rights leader Angela Davis speaks at Bovard*. Daily Trojan, 23 February, 2015.

Foucault, M. (1977). *Discipline and punish: The birth of the prison*. Penguin.

Goyes, D. (2019). *Southern green criminology: A science to end ecological discrimination*. Emerald.

Goyes, D., Abaibira, M., Baicue, P., Cuchimba, A., Nenetofe, D., Sollund, R., South, N., & Wyatt, T. (2021). Southern green cultural criminology and environmental crime prevention: Representations of nature within four Colombian indigenous communities. *Critical Criminology, 29*, 469–485.

Graham, M. (2008). Some thoughts about the philosophical underpinnings of Aboriginal worldviews. *Australian Humanities Review, 45*, 181–194. http://www.australianhumanitiesreview.org/archive/Issue-November-2008/graham.html. Accessed 21 November 2016.

Huanca, Y. K. A. (2019). Non-western epistemology and the understanding of the *Pachamama* (environment) within the world(s) of the *Aymara* identity. *International Journal for Crime, Justice and Social Democracy, 8*(3), 6–22.

Johnston, E. (1991). Royal commission into aboriginal deaths in custody, National Report. Volume 2: *The Underlying Issues Which Explain the Disproportionate Number of Aboriginal People in Custody.* Australian Government Publishing Service.

Johnston, I. (2015). Let them feed him biscuits: Doing fieldwork in Fiji with the family. *Forum Qualitative Sozialforschung/Forum: Qualitative Social Research, 16*(1), Art. 17. http://www.qualitative-research.net/index.php/fqs/article/view/2220. Accessed 21 November 2016.

Kidman, J. (2020). Whither decolonisation? Indigenous scholars and the problem of inclusion in the neoliberal university. *Journal of Sociology, 56*(2), 247–262.

Kipling, R. (1899). The White Man's Burden: The United States & The Philippine Islands, 1899. *Rudyard Kipling's Verse: Definitive Edition.* Doubleday, 1929.

Kniest, P. (2021). The changing university workforce landscape: A response to regulation, funding or the emergence of the enterprise university. *Advocate, 28*(1), 22–23.

Monod de Froideville, S., & Bowling, R. (2022). Te Awa Tupua: An Exemplary Environmental Law? In J. Gacek, & R. Jochelson (Eds.). *Green criminology and the law.* Palgrave Macmillan.

Moosavi, L. (2020). The decolonial bandwagon and the dangers of intellectual decolonisation. *International Review of Sociology.* https://doi.org/10.1080/03906701.2020.1776919

Robyn, L. (2002). Indigenous knowledge and technology. *American Indian Quarterly, 26*(2), 198–220.

Said, E. (1994). *Representations of the Intellectual – The 1993 Reith Lectures.* Vintage.

Tauri, J. (1999). Explaining recent innovations in New Zealand's criminal justice system: Empowering Maori or biculturalising the state? *Australian and New Zealand Journal of Criminology, 32*(2), 153–167.

United Nations Economic and Social Council. (2010). Study on the Need to Recognize and Respect the Rights of Mother Earth. New York: Permanent Forum on Indigenous Issues, Ninth session, 19–30 April, United Nations.

Walter, M., & Suina, M. (2019). Indigenous data, indigenous methodologies and indigenous data sovereignty. *International Journal of Social Research Methodology, 22*(3), 233–243. https://doi.org/10.1080/13645579.2018.1531228

Walter, M., Kukutai, T., Carroll, S., & Lonebear, D. (Eds.). (2021). *Indigenous data sovereignty and policy.* Routledge.

Watson, I. (2015). *Aboriginal peoples, colonialism and international law: Raw law*. Routledge.
White, R. (2001). Criminology for sale: Institutional change and intellectual field. *Current Issues in Criminal Justice, 13*(2), 127–142.
White, R. (2008). *Crimes against nature*. Routledge.
White, R. (2013). *Youth gangs, violence and social respect: Exploring the nature of provocations and punchups*. Palgrave Macmillan.
White, R. (2016). The four ways of eco-global criminology. *International Journal for Crime, Justice and Social Democracy, 6*(1), 8–22.
White, R. (2022). *Theorising green criminology: Selected essays*. Routledge.
White, R., & Graham, H. (2015). Greening justice: Examining the interfaces of criminal, social and ecological justice. *British Journal of Criminology, 55*, 845–865.
Wright Mills, C. (1959). *The sociological imagination.* Oxford University Press.

Index

D. R. Goyes (ed.), *Green Crime in the Global South*, Palgrave Studies in Green Criminology, https://doi.org/10.1007/978-3-031-27754-2

SPRINGER NATURE

GPSR Compliance

The European Union's (EU) General Product Safety Regulation (GPSR) is a set of rules that requires consumer products to be safe and our obligations to ensure this.

If you have any concerns about our products, you can contact us on ProductSafety@springernature.com

In case Publisher is established outside the EU, the EU authorized representative is:

Springer Nature Customer Service Center GmbH
Europaplatz 3
69115 Heidelberg, Germany

The manufacturer's authorised representative in the EU is Springer Nature Customer Service Centre GmbH, Europaplatz 3, 69115 Heidelberg, Germany. If you have any concerns regarding our products, please contact ProductSafety@springernature.com

Printed and bound by CPI Group (UK) Ltd, Croydon, CR0 4YY

01/10/2026

02225851-0001